THE ILLUSTRATED GUIDE TO
DUCKS AND GEESE
AND OTHER DOMESTIC FOWL

How to choose them – How to keep them

BLOOMSBURY

LONDON · BERLIN · NEW YORK · SYDNEY

First published in 2012

Copyright © 2012 text and illustrations by Celia Lewis

The right of Celia Lewis to be identified as the author of this work has been asserted by her in accordance with the Copyright, Designs and Patents Act 1988.

Bloomsbury Publishing Plc, 50 Bedford Square, London WC1B 3DP
Bloomsbury USA, 175 Fifth Avenue, New York, NY 10010

www.bloomsbury.com
www.bloomsburyusa.com

Bloomsbury Publishing, London, New York, Berlin and Sydney

A CIP catalogue record for this book is available from the British Library
Library of Congress Cataloging-in-Publication Data has been applied for

Commissioning Editor: Julie Bailey
Design by Julie Dando at Fluke Art

UK ISBN (print) 978-1-4081-5264-5
US ISBN (print) 978-1-60819-975-4

Printed in China by C&C Offset Printing Co Ltd.

This book is produced using paper that is made from wood grown in managed sustainable forests. It is natural, renewable and recyclable. The logging and manufacturing processes conform to the environmental regulations of the country of origin.

10 9 8 7 6 5 4 3 2 1

Contents

Introduction

By far the largest group of domestic fowl is of course chickens, but records show that ducks and geese have also been domesticated for thousands of years. Peafowl, quail, guineafowl and turkeys have also been domesticated over the years and provide us with a wide variety of birds to choose from – truly something for everyone, whether you are blessed with rolling acres or simply have a small backyard.

It is possible to keep peafowl and guineafowl in fenced runs but their wild spirits mean they will be far happier running free range and roosting high in trees. These two species also have very loud voices and are not shy of using them, so neighbours must be considered.

Geese will need a large area of grassland as well as access to water. They can be aggressive, particularly during the breeding season. As a child my grandmother kept geese where she lived in the New Forest; every morning they walked in a tidy line into the forest with Hillary, the gander, fussing round them. Woe betide you if you met them on their way as Hillary would try to see you off, rushing at you with his head down and hissing – a terrifying sight for a small child with bare skinny legs, and I have kept a healthy respect for geese ever since.

An angry African gander

The goose hisses,
but does not bite.
Danish proverb

Pearl guineafowl

Turkeys can be kept enclosed but will obviously need a large house with special perches and will do far better if they have plenty of space to roam, although they too will prefer to roost in trees if given the chance. Ducks, not surprisingly, require water but some will be happy with the tiniest of ponds – even a child's paddling pool will meet the needs of a small flock. Quail can live in an even smaller area.

Take into account that you will probably need to tend your birds twice a day, either to let them out in the morning and shut them up again at night, or to feed them. Feeding is the ideal time for 'lookering', or checking that all is well with your flock – cast your eye briefly over each bird and you will soon learn to spot that something is not right.

The birds illustrated on the profile pages are all actual birds and, although typical of their breed, are not necessarily of show standard – if you're thinking of showing your birds you should consult the breed standards for your chosen breed.

Pilgrim goose

Feathers

Feathers are made of keratin, a fibrous protein that also makes up hair, hoof, nails and horn. Ducks and geese have different feather types: down, to keep them insulated and warm; and flight feathers found in the wings.

Guineafowl feathers

During the moult birds cast off their outer feathers but retain the down, which regenerates naturally throughout the year. Geese moult just once a year, goose and gander together, when the goslings are 2 or 3 weeks old. The feathers re-grow in a matter of 3–4 weeks, by which time the goslings are 6–8 weeks old. Male ducks moult twice a year, gaining their most colourful plumage in time for the breeding season, then shedding this to go into 'eclipse' during the summer months – this makes them less eye-catching to predators. The ducks retain their camouflage colouring all year round but moult when their ducklings are grown.

Peafowl, guineafowl, turkeys and quail also moult annually in the autumn.

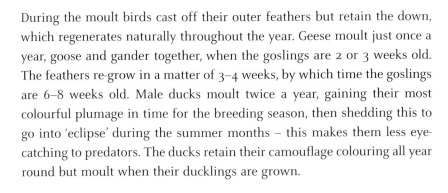

Toulouse goose feathers

Although feathers are light, a bird's plumage weighs two or three times more than its skeleton, since many bones are hollow and contain air sacs.

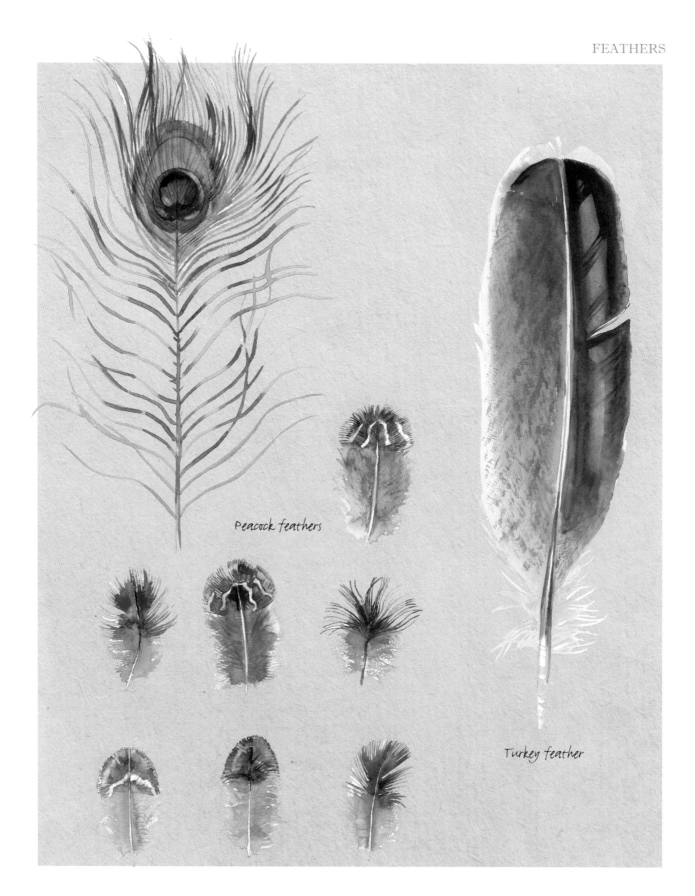

Peacock feathers

Turkey feather

Roman goose preening

Wing clipping is a means of containing your birds. Small breeds of duck or light geese will tend to fly, and a way of unbalancing them and making this difficult is to carefully clip off 5 cm (2 in) or so of the flight feathers of one wing (not both wings as then the bird would still be balanced and able to fly). This should only be done when the feathers are fully formed and may not be necessary for larger breeds as they either cannot – or choose not to – fly.

Feathers have all sorts of uses, from fishing flies to quills – the flight feathers of the goose make the best quill pens. Duck and goose down is used in duvets, pillows, sleeping bags and jackets and makes marvellously light and warm products thanks to its insulating properties. Peacock and turkey feathers appear on hats and masks and even earrings, fans and handbags.

The fletchings or feathers on a longbow arrow are made from turkey feathers.

How to make a quill pen

Traditionally goose flight feathers were used to make quills as they were considered the strongest, but in fact any large feather with a shaft of at least 0.5 cm (0.25 in) can be used.

The traditional way to prepare a quill was to stick the feather in a tin of very hot sand and leave it there until the sand went cold – the shaft would then have become opaque and ready to be worked.

A simpler way is to dip it in boiling water to soften it, rather as you would soften a toenail in hot water.

1. Select your feather and hold it in your hand as if to write – remove all the plume that touches your hand or it would be uncomfortable to hold.
2. Dip the shaft in boiling water for 5 minutes.
3. Cut away the tip at an angle of 45 degrees.
4. Clean out the membrane.
5. Cut a scoop from the underside.
6. Shape the nib to match on each side.
7. Place the nib face down on a hard surface and cut a slit in the middle, approximately 10 cm (4 in) long.
8. To 'nib', rest the underside on a hard surface and thin the tip by scraping.
9. Sand and clean.
10. Try your pen and admire your handiwork.

Oh, nature's noblest gift, my grey goose quill,
slave of my thoughts, obedient to my will,
Torn from the parent bird to form a pen,
That mighty instrument of little men.

Lord Byron

A quill pen

Eggs

Eggs are one of the most nutritious foods available and provide all the essential amino acids and minerals required by the human body. They contain vitamins A, B and D and supply a complete protein that is of a higher quality than all other food proteins.

Eggs consist of three parts: the shell, which is made of calcite (a crystalline form of calcium carbonate); the albumen or white, which is made up of water and protein; and the yolk, which is the most nutritionally valuable part of the egg. The colour of the yolk is directly associated with the bird's diet. A very pale yolk can indicate underfeeding or lack of greens, whereas a bird that has access to plenty of vegetable matter will produce eggs with a much darker yolk.

A fresh egg will sink in water, a stale one will float.

Guineafowl eggs can vary slightly in size and colour but are mostly light beige to off-white with occasional speckles. The shell is particularly thick in comparison to hen's eggs and the narrow end more conical. This particular shape of egg is characteristic of seabirds that nest on ledges and prevents the eggs from rolling off – for the guineafowl (and also peahen and turkey) the shape enables more eggs to be laid in a small space as they fit neatly together. Guineafowl tend to lay in the middle of the day and a good layer will produce up to 30 eggs in a clutch.

Peahens are more likely to lay late in the afternoon, with 7–10 eggs in a clutch. If the eggs are collected regularly they will lay up to 20 or 25 pale creamy buff eggs in a year.

Turkeys lay between 50 and 100 eggs a year weighing up to 100 g (3½ oz). The eggs are mainly cream-coloured with brown speckles and about twice the size of a normal hen's egg but with a tougher shell and membrane – they taste similar and can be used in exactly the same way.

Geese lay the largest eggs, measuring up to 10 cm (4 in), but are seasonal layers – unlike ducks, which can lay all winter in some cases. Quail lay the smallest eggs, but that is not to say the taste is inferior – in fact they're considered a delicacy and usually served hard-boiled with celery salt. Of all the domestic eggs produced, ducks' eggs are considered the best for baking.

If the stone fall upon the egg, alas for the egg!
If the egg fall upon the stone, alas for the egg!
Arabic proverb

Goose egg

Duck egg

Cayuga duck egg

Coturnix quail egg

Guineafowl egg

Peacock egg

Turkey egg

Breeding

You can acquire eggs from many different sources if you do not have your own. They can be sourced from poultry shows, magazines or breed clubs, and most eggs can even be safely delivered by post.

Rouen Clair duck with ducklings

Using a broody

If your birds don't go broody themselves and you want to breed then a broody hen will do all the work for you. A hen may well be more compliant than a duck or goose, and certain breeds go broody more readily than others and make better mothers. Wyandotte, Orpington and Light Sussex are all very good and frequent broodies. Firstly make sure your hen really is broody. If possible move her (tactfully, at night) to a nest box well away from other birds. Set her on some old or china eggs (even golf balls would do) to make sure she is settled.

skåne goose

You have a week or so to collect your own fertile eggs or acquire some. Choose clean, well-shaped eggs and keep them in a cool place – they should not be warm, about 12°C (55°F) is fine. Do not wash or wipe them – duck eggs tend to be dirty, so try and choose the cleanest ones; if you have to clean them do so carefully with a nail brush. Place them pointed end down in an egg box and alter their position once a day. An easy way to do this is to prop the box up on one side and swap sides each day. They will keep for at least 14 days but hatchability goes down after about 10 days; those that have come by post should be no more than 7 days old. Eggs that arrive by post should be given at least 12 hours to rest before being introduced to the broody.

Once you have collected your clutch, which could be up to 12 duck or guineafowl eggs or 9 turkey or peacock eggs for a full-sized hen, remove the false eggs and pop the hatching eggs under the broody, being careful not to upset her – this is best done at night.

She will now settle in for the number of days it takes for the eggs to hatch, carefully turning them several times a day. She will probably only leave her nest once a day to drink, eat and defecate – don't worry if you do not see her get off, just try not to disturb her. Leave her food – she will be happy with a couple of handfuls of wheat – and make sure she has fresh water.

In her natural state, the hen would dampen her breast in the dew on the grass when she gets off her nest to eat and drink – if she's confined to a shed or has no outside access it's a good idea to sprinkle the eggs with lukewarm water very carefully on the last few days before hatching is due (hatchlings formed but dead in the shell when the hatching date is reached is a sign of lack of humidity).

The eggs will start to 'pip' – this is when the hatchling begins to break through the shell using its egg tooth, a horny growth on the end of its beak. Once it has made a small hole it will rest for up to 8 hours, and at this point you can often hear the hatchlings already cheeping from inside their shell. The hatchling will then continue the laborious task of breaking free. The mother will wait until she's sure that all the hatchlings have hatched and dried before bringing them out – they can happily survive for 36 hours with no food or water, living off the remaining yolk in their stomachs. If after 36 hours you suspect there are still unhatched eggs then it is best to remove them. Have ready a low container of hatchling crumbs and a water container that the hatchlings can reach into but not drown in – the addition of a few stones will help to make it safe. Don't forget food for the broody herself – you can give her mixed corn and she will break it up for the hatchlings and encourage them to eat the crumbs.

Pipping

Hatching

Using an incubator

Without a broody you can still rear hatchlings by using an incubator – the results are never quite as good and you should expect the occasional failure. Incubators come in all sorts and sizes, from tiny models that will require hand turning of the eggs several times a day, to medium-sized that will take 20–40 eggs, depending on size, and do the turning automatically, up to enormous commercial cabinets that may take several hundred eggs. There are two types of incubators – forced-air machines that have a fan built in to circulate the air, and still-air incubators with no fan. What they all have in common is that the humidity and temperature must be correct – it should be 37.7°C (100°F) for forced-air and 39.4°C (103°F) for still-air incubators.

Machines will all be slightly different so read the instructions that come with yours to find out the correct temperature and exactly how much and where the water for humidity should be added. During incubation the water will evaporate and need to be topped up – always add warm water as close as possible to the temperature in the incubator. Each machine will have a ventilation hole – follow the instructions for guidance on how open or closed it should be.

After 8–10 days you can 'candle' the eggs to see whether they are fertile. This involves holding each egg in front of a bright light. Special candling lamps can be bought but home-made versions work just as well. You need a bulb or torch inside a box with an egg-shaped hole on the top. The egg is held over the hole so that the light shines through and you can see inside. This is best done in a darkened room. At 7–10 days a fertile egg will have blood vessels that look rather like leggy spiders, and an obvious air sac at the broad end. At 14 days any infertile eggs will appear clear and should be removed, and those growing correctly will have a large shadow and enlarged air sac.

just hatched

up on its feet

Call ducklings

For the last 3 days before hatching remove the machine from its cradle or stop turning and increase the humidity slightly. Try to resist opening the machine when hatching starts (this will be hard) to maintain humidity. Pipping will start a couple of days before hatching is due, when the hatchling will break through the shell with its special egg tooth. Remember that once pipped the hatchling may rest for at least 8 hours before continuing to hatch and with larger birds the whole process may take up to 48 hours. Never help a hatchling from its shell – it will hatch on its own or there will be a reason why it does not.

If you're unlucky enough to have a power cut while incubating all may not be lost. There are various methods of keeping the eggs warm. Firstly cover the incubator with blankets or find a box large enough to cover it to form a makeshift frame and set night-light candles in jam jars inside. This should be sufficient to maintain warmth until the power returns. Embryos can survive at slightly lower temperatures for up to 18 hours, although this may mean the hatch will be a day later.

The hatchlings will rest for a while when they have finally broken free of the shell but they will soon be on their feet. As soon as they have all dried and fluffed up, remove them to an already prepared rearing pen with a heat lamp.

Blue swedish duckling

The rearing pen

If you have room, create a circular enclosure with cardboard or other flexible material so that there are no corners for hatchlings to get trapped in. Hang the lamp slightly to one side and at a height such that the temperature is 32°C (90°F) at ground level. The hatchlings will also need shallow containers for chick crumbs and special chick drinkers. When introducing the hatchlings to their new home dip their beaks carefully in the water and place them under the lamp. They do not need to eat or drink on their first day but make sure they have found the food and water on the following day. You will be able to see that the temperature is correct – the hatchlings will spread right out away from the lamp if it is too hot and crowd together underneath if too cool. Each week raise the lamp a little until by week three the temperature is no more than 23°C (75°F). During summer the hatchlings should no longer need a heat source once they are 5 weeks old – but keep it on for another week or so in cold weather.

Typical heat lamp

When your hatch is successfully over make sure to remove all shell debris from the incubator and rinse out the inside, making sure it's totally dry before closing it up.

Day-old hatchlings can successfully be introduced to a broody who has been sitting on sham eggs. Do this at night in as quiet a way as possible. Gently remove a couple of the sham eggs and pop a couple of day-olds in their place. Wait to see if the hen will accept them before continuing to remove the eggs and replace them with hatchlings. Broodies are less likely to accept hatchlings that are more than 2 days old but if she has successfully adopted a few and then more of your incubated eggs hatch she may not notice if you slip them under her at night.

Broody hen with ducklings

DUCKS

Introduction

All ducks are descended from the wild Mallard, with the exception of the Muscovy which claims the wood ducks of Latin America as its ancestor.

Ducks are hardy birds, hardier than chickens thanks to a second layer of feathers known as down close to their skin and an extra layer of fat needed to keep them warm in cold water. Feather care is very important and ducks will spend a good deal of their time keeping their feathers waterproof by oiling them from the preen gland near the tail. They need to be able to 'duck and dive', which consists of ducking their heads in water in order to throw it over their backs, rolling around and beating their wings on the water, so if a suitable pond or stream is not available they must have a container large enough to do this – a garden sprinkler will be greatly appreciated in hot weather. If the run is muddy and there is a lack of clean water, the birds can suffer from 'wet feather' (see page 155). This is a serious problem that is exactly what it sounds like – the feathers lose waterproofing and look constantly wet and the bird will be miserable and lose condition.

Ducks' feet have no nerves or blood vessels. This means ducks never feel the cold, even if they swim in icy cold water.

Ducks are designed to swim, their distinctive waddle caused by the fact that their legs are set well back on their bodies – excellent for swimming in water but awkward on land.

Certain breeds lay more eggs than hens and the eggs are lower in cholesterol.

Housing and fencing

Although ducks are hardier than chickens they still need a shed to protect them at night. Unlike hens they do not perch, so any garden shed or similar will be sufficient as long as the floor is dry. Make sure there is adequate ventilation and site the house in a corner of the run so that they can easily be driven in if they don't learn to go to bed by themselves.

Blue swedish preening

You can choose to clean out your house every day or once a week, but the simplest option is to go for deep litter. With this method bedding doesn't have to be constantly changed but can be maintained by adding fresh material when needed to build up a deep warm bed – mucking out every few months. The best bedding is wood shavings or sawdust, but hay and straw can also be used, although they tend to get wet more quickly and will need cleaning out more frequently.

Take like a duck to water. To learn how to do something very quickly and to enjoy doing it.

Ducks can be kept in by a 1 metre (3 foot) fence, but this will not keep foxes out and they are a formidable foe. In order to be safe your ducks will require chicken-wire fencing at least 2.1 metres (7 feet) high, preferably either dug into the ground a foot or so all round or with wire laid flat on the ground for 50 cm (18 in) outside the fence to deter animals from digging underneath. Most domestic ducks either don't or can't fly but if you have an active breed that does, clipping the flight feathers of one wing to unbalance the bird will stop them. Don't make the mistake of clipping both wings as the birds will still be able to fly!

Feeding

Ducks are omnivorous and as such have a varied diet that includes plants, snails, slugs and worms from above or below the water. Domestic ducks will need additional feeding, particularly if they are laying, and commercially produced pellets are carefully balanced to provide total nutritional needs.

A fully grown duck laying or resting will need 170 g (6–7 oz) of food daily, made up half and half of duck pellets and mixed corn. As with chickens it is best to feed twice a day, giving the grain in the afternoon. If your ducks are running free all day, regular feeding also has the advantage of bringing them back in the evening, and if you have a small fox-proof run then feeding them inside it will make shutting them up for the night considerably easier, particularly in midsummer with the long light evenings.

All birds need grit to grind the food in the gizzard and although your ducks may be able to find a good deal for themselves, commercially produced grit and oyster shell should always be available to them.

Ducklings will need duck starter crumbs up to 3 weeks of age followed by duck grower pellets – don't feed chick crumbs if they contain any sort of medication (they often contain an anti-coccidia drug) as it's easy for the ducklings to eat too much and overdose themselves. Ducks that are expected to start laying at around 24 weeks should be fed layers pellets from 16 weeks of age.

Store the food in feed bins or dustbins to prevent mice and rats from helping themselves and also to keep it dry.

Feeding is not an exact science and obviously very large ducks, or those who are prolific layers, will need considerably more food than a little Call duck. Feed the pellets in containers – old roasting pans are ideal. You'll know if you are feeding too much if there is anything left after about 20 minutes. Pellets left lying around encourage rats and quickly go mouldy. Wheat can be fed under water, which has the advantage of preventing other wild birds such as jackdaws from helping themselves.

A duck will not always dabble
in the same gutter.
Danish proverb

Water

If you don't have a pond or stream, a small paddling pool or ornamental garden pond will be adequate but it must be easy to empty and refill. A small solid-sided artificial pond would be fine as long as it can be topped up and drained. Make sure the ducks can get out either by using a ramp or piling some bricks inside near the edge. You could also create a pond by digging a hole and using a pond liner. Whatever method you use it's a good idea to surround the pond with either concrete or shingle or it will very soon become a muddy mess.

Abacot Ranger

Some birds avoid water, the duck seeks it.
Hausan

Breeding

You will of course need a drake in order to have fertilised eggs. Some ducks and drakes look very similar, but the drake will always have the distinctive curled 'sex feather' at the base of his tail. Ducks mate on water and a duck's eggs may not be fertile if adequate water is not available. If possible wait until your birds are 2 years old before breeding from them, when their eggs will have reached full size. If you have two different breeds, or indeed two drakes, in order to be sure the offspring comes from your chosen one, the birds must be separated for at least 14 days (21 days would be safer) before mating. Some ducks are excellent mothers but many are not, and a broody hen or incubator may well produce better results.

Only use perfect eggs for hatching. Mark them with the date you collect them and if possible use clean eggs – carefully wipe any dirt off those that aren't clean. Store the eggs in a cool room and turn them daily until you have collected your clutch and are ready to 'set' them. Try not to leave them longer than 7 days if you're going to put them in an incubator, although they will be alright for a few more days if a broody hen is going to do the work for you. If they have come by post let them settle for 24 hours.

**Collective nouns for ducks:
on water – a raft, badling or
paddling of ducks; in flight –
a skein, team or string of ducks.**

Muscovy duck with ducklings

A domestic duck's incubation period is 28–30 days, although a Call duck will hatch in 26–27 and the Muscovy a very long 33–35 days.

Follow the method outlined on page 15.

Ducklings are messy creatures and will spill their food and water – if they're in a small space it's possible to create brood boxes with grid bases off the floor. They will need a layer of newspaper at first in order to keep them clean and dry. If there is plenty of room then newspaper and straw should be adequate if it is changed on a regular basis.

If it walks like a duck, quacks like a duck, looks like a duck, it must be a duck.
Italian proverb

Campbell duck with ducklings

Parts of a duck

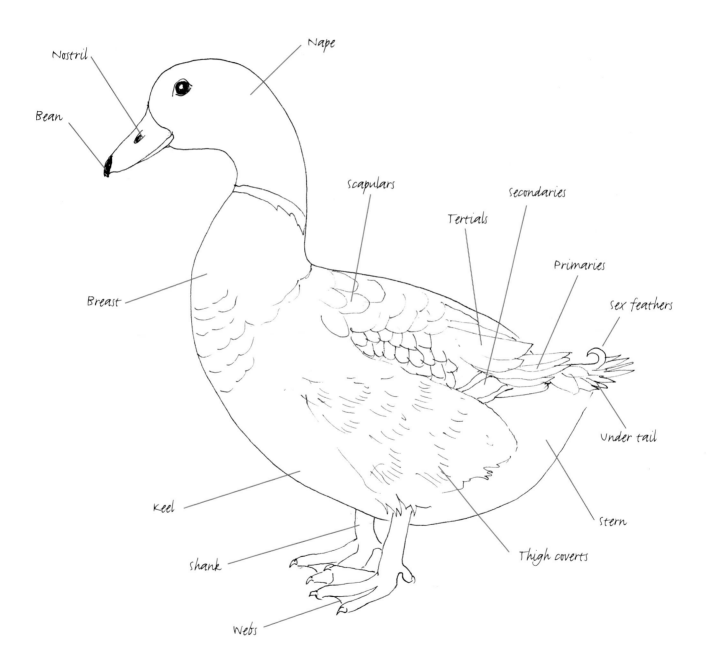

Nostril

Bean

Nape

Bean

Breast

Scapulars

Tertials

Secondaries

Primaries

Sex feathers

Under tail

Stern

Keel

Thigh coverts

Shank

Webs

Abacot Ranger

An attractive all-rounder

Along with many other breeds the Abacot Ranger (known in the USA as the Hooded Ranger thanks to the fact that both duck and drake have well-defined hoods) has Indian Runner in its blood. Oscar Gray from Colchester is credited with developing the breed around 1920 by crossing a Khaki Campbell 'sport' with a white Indian Runner drake. What he produced was a placid bird, easy to tame, similar to a Khaki Campbell in shape (although larger) with attractive plumage. This is a long-lived breed – birds can live to 10 years of age.

The Abacot Ranger has several advantages, one being that ducklings can be sexed from 8 weeks of age by the colour of their bills. The duck's is dark slate and the drake olive green. The fact that they rarely, if ever, fly is also in their favour as they can be kept in by a low fence as long as predators are kept out. They are also excellent layers of large white eggs.

Abacot Ranger ducklings

Abacot Ranger duck

ORIGIN	CLASS	EGG COLOUR	NO. OF EGGS PER YEAR	
UK	Light	White	180–200	
	ADULT WEIGHT	BILL COLOUR	LEG COLOUR	EYE COLOUR
DRAKE	2.3–2.5 kg 5–5½ lb	Olive-green with black bean	Orange	Dark brown
DUCK	2.0–2.3 kg 4½–5 lb	Dark slate-tinged green	Orange	Dark brown

Abacot Ranger drake

Aylesbury

The traditional meat-producing bird

Not surprisingly this breed hails from Aylesbury in Buckinghamshire, where it was bred for the London market. Once known as the White English thanks to its white skin, it has a characteristic pink bill. Apart from being bred for meat the feathers were used for quilts and pillows.

In the eighteenth century the occupants of nearly every cottage in Aylesbury would have bred ducks, keeping them inside their cottages in wooden crates. They were walked to London – a distance of nearly 40 miles – in large flocks, staying at inns along the way that provided special enclosures. In order to protect their feet on this marathon trek they were walked through a kind of sticky tarry substance and then through sawdust every morning.

This is a rather greedy breed – therefore easy to fatten – that can become overweight. This fact means that they rarely fly and also that they are rather clumsy mothers; broody hens are usually more successful in hatching eggs.

> **Lovely weather for ducks.** *Cliché.* A greeting meaning that unpleasant rainy weather must be good for something.

Aylesbury drake

ORIGIN	CLASS	EGG COLOUR	NO. OF EGGS PER YEAR	
UK	Heavy	Greenish-white	40–100	

	ADULT WEIGHT	BILL COLOUR	LEG COLOUR	EYE COLOUR
DRAKE	4.5–5.4 kg 10–12 lb	Pink	Orange	Blue
DUCK	4.1–5.0 kg 9–11 lb	Pink	Orange	Blue

Aylesbury duck

Aylesbury duckling

Bali

Elegance is the word

A duck indigenous to the island of Bali, east
of Java, this breed was imported into the
UK in the 1920s. Although the original blood line may have
been lost more recent birds have been created by crossing Crested
ducks with Indian Runners. This is an ancient breed – Asian temple
carvings dating back 2,000 years show upright ducks with crests.

Birds are mostly found completely white, although brown is the
more common colouring in their homeland – any colour variation is
acceptable.

This beautiful elegant bird is quite capable of flying if caught
off-guard and frightened, so either a high fence
will be required or one wing will have
to be clipped to contain them.

In common with Indian Runners,
Bali are difficult to breed, with a high
proportion of dead-in-shell ducklings.

Bali drake

Bali duck

ORIGIN	CLASS	EGG COLOUR	NO. OF EGGS PER YEAR	
South Asia	Light	Blue-green to white	100–200	
	ADULT WEIGHT	BILL COLOUR	LEG COLOUR	EYE COLOUR
DRAKE	2.3 kg 5 lb	Orange-yellow	Orange	Blue
DUCK	1.8 kg 4 lb	Orange-yellow	Orange	Blue

Bali duck

Black East Indian

A beautiful duck with a handy appetite

This little bantam duck is similar in colouring to the Cayuga, with gorgeous glossy plumage that shines beetle-green in sunlight. It is thought to have been imported into the UK from Brazil by the Earl of Derby in 1850, though references exist to show it was already in the country in 1831.

Best kept as a pair or trio, the duck will lay two clutches a year and will only go broody if the eggs are not removed. These are good and willing flyers and will need to have a wing clipped if they are to live in any sort of enclosure.

The drakes retain their dark colouring but ducks often develop pale patches with age. They have the reputation that they are especially fond of slugs.

Black East Indian drake

ORIGIN	CLASS		EGG COLOUR	NO. OF EGGS PER YEAR	
Unknown	Bantam		Almost black	80–120	
	ADULT WEIGHT	**BILL COLOUR**	**LEG COLOUR**		**EYE COLOUR**
DRAKE	0.9 kg 2 lb	Black	Black		Dark brown
DUCK	0.8 kg 1¾ lb	Black	Black		Dark brown

Black East Indian duck

Black East Indian duckling

Blue Swedish

A large possessor of unusual-coloured plumage

Blue ducks have been known in Europe since the nineteenth century but have been around longer as they appear in seventeenth century Dutch paintings. These large ducks with a smart white bib are genetically black with a single blue dilution and come in three colours: black, blue and pale blue-white. Blacks mated to blacks will produce black ducklings, pale blue to pale blue produce all pale blue, but if you breed a black drake with a pale blue duck you will achieve a beautiful dark blue colour.

This is a hardy breed that produces a fine plump carcass by 16 weeks and has a slightly more upright stance than some other heavy breeds. This duck rarely flies so it is easily kept in, and it would make a good all-round choice for the backyarder.

Blue swedish drake

Blue swedish duck

Duckling

ORIGIN	CLASS	EGG COLOUR	NO. OF EGGS PER YEAR	
Europe	Heavy	White	100–150	
	ADULT WEIGHT	BILL COLOUR	LEG COLOUR	EYE COLOUR
DRAKE	3.6 kg 8 lb	Blue preferred to green	Orange-brown	Brown
DUCK	3.2 kg 7 lb	Slate blue	Blue-brown	Brown

Blue Swedish drake

Call

The little duck available in any colour

This bantam breed of duck was once known as Decoy, but the name was changed to Call in 1870, apparently from the Dutch 'kooi' meaning cage or trap. They were used to entice wild birds onto a pond or lake to be shot. They have exceptionally loud voices and enjoy swimming, meaning that they are really only suitable if you have a fair-sized pond and no close neighbours.

Their round faces and short bill make them look rather cute but they are feisty birds, easy to tame and excellent flyers – your run may need a net over it to contain them. Birds will sit and make good mothers but only lay one clutch a year. When they hatch the ducklings are tiny and therefore easy prey for aerial predators as well as ground-based ones.

Call ducks come in nearly every colourway imaginable – 17 colour varieties are recognised by the British Wildfowl Association.

An ugly duckling. Someone that is ugly and not successful when they are young or new, but who develops into someone beautiful and successful.

Call ducks and drakes

ORIGIN	CLASS	EGG COLOUR	NO. OF EGGS PER YEAR		
Holland via Asia	Bantam	White	Up to 30		
	ADULT WEIGHT	BILL COLOUR	LEG COLOUR		EYE COLOUR
DRAKE	0.6–0.7 kg 1¼–1½ lb	Varied	Varied		Varied
DUCK	0.5–0.6 kg 1–1¼ lb	Varied	Varied		Varied

Call duck

Call ducklings

Campbell

The perfect garden ornament

Named after its original breeder, a Mrs Adele Campbell of Gloucestershire, who wanted to produce a good laying table bird that was reluctant to fly and rarely went broody – and this is exactly what she produced. She crossed White Indian Runners with Rouen and added in a bit of Mallard to increase hardiness. The result was a khaki-coloured duck that in some ways behaved more like a hen – it only rarely flies and is quite happy with a very small amount of water. A large washing-up bowl will suffice as long as the birds can get their heads under the water – weighting it down with a brick or setting it in a trench will stop the ducks constantly tipping it over.

Campbell drake

As well as the normal khaki colour, Campbells also come in white and dark khaki, although the latter is becoming increasingly rare. This breed has acquired a reputation for being unreliable sitters on the occasion that a duck does go broody – abandoning the nest before the eggs are ready to hatch. There is an advantage for the breeder, however, in mating a khaki drake with a dark Campbell duck which then produces sex-linked ducklings, the females being khaki and the males much darker.

A handsome and rewarding choice for the smallholder.

ORIGIN	CLASS		EGG COLOUR	NO. OF EGGS PER YEAR	
UK	Light		White	300–350	
	ADULT WEIGHT	**BILL COLOUR**	**LEG COLOUR**	**EYE COLOUR**	
DRAKE	2.3–2.5 kg 5–5½ lb	Dark: bluish-green, Khaki: greenish-blue White: orange	Dark: dark orange Khaki: dark orange White: orange	Dark: brown Khaki: brown White: grey-blue	
DUCK	2.0–2.3 kg 4½–5 lb	Dark: lead Khaki: dark slate White: orange	Dark: dark orange Khaki: dark orange White: orange	Dark: brown Khaki: brown White: grey-blue	

Campbell duck

Cayuga

An exotic addition to any backyard

A native American tribe called Cayuga gave their name to Lake Cayuga in New York State and in turn the duck was named after the area from whence it came. Probably related to a wild American black duck it is also possible that Mallard featured in its heritage. Whatever its ancestry, this is a stunning-looking bird with black feathers that have such a sheen and lustre they appear iridescent green, particularly in sunlight. It is preferred that the sheen remains green and does not have purple undertones.

Originally reared for their meat, they lost popularity when the white Pekin appeared, with its more attractive white skin. They are now usually kept for their beauty alone, although they do still make excellent table birds.

Cayuga drake

The eggs of a Cayuga appear almost black, the colour being produced by a black pigment that can be scratched off, revealing that the shell is white underneath. As more eggs are laid the colour can change to green or white. The duck also loses some of her dark colouring with age – developing patches of white – although the drakes retain their dark looks.

Cayugas are ideal ducks for smallholders. They seldom fly except when young, but do need good access to water, particularly in summer as their dark colouring causes them to feel the heat more than most.

ORIGIN	CLASS	EGG COLOUR	NO. OF EGGS PER YEAR	
USA	Heavy	Dark brown to almost black	80–100	

	ADULT WEIGHT	BILL COLOUR	LEG COLOUR	EYE COLOUR
DRAKE	3.6 kg 8 lb	Black	Black as possible	Dark brown
DUCK	3.2 kg 7 lb	Black	Black as possible	Dark brown

Cayuga drake

Cayuga duckling

Crested

An unusual character with a powder-puff on its head

Crested ducks have been known since the early seventeenth century in Europe, when they appeared in Dutch Master paintings, but it is thought that they originated in Asia along with the Bali.

This is a beautiful duck but difficult to breed, because the crest is the result of a lethal gene that causes a gap in the skull through which fatty tissue protrudes and causes the feathers to produce the crest. Many ducklings fail to hatch and those that do may suffer other abnormalities – not a breed for the amateur.

Found most commonly in their white form they can also come in any other colour. The crest or pouf is the most important part of their make-up in the show ring and should be in the centre of the head, not over to one side or too far back or forward, and be neatly circular.

There is also a miniature version of the Crested duck, developed in the late 1980s and standardised by the Poultry Club of Great Britain in 1997.

Crested drake

Crested duckling

ORIGIN	CLASS	EGG COLOUR	NO. OF EGGS PER YEAR	
Unknown	Light	White	100–130	
	ADULT WEIGHT	BILL COLOUR	LEG COLOUR	EYE COLOUR
DRAKE	3.2 kg 7 lb	Orange-yellow	Orange	Blue
DUCK	2.7 kg 6 lb	Orange-yellow	Orange	Blue

Crested duck

Hook Bill

A duck with a Roman nose?

A very old breed whose ancestry is uncertain but is probably of Indian origin. Rare in the UK and USA, this breed is very popular in Holland – traditionally the birds were let out onto the canals every day and returned home on their own every evening. They would, however, make an excellent choice for the smallholder, being easy to tame, energetic foragers, and not too big or heavy. They can easily be deterred from flying by clipping one wing.

Hook Bill duck

They are average layers of pretty blue eggs which for some reason do not seem to do well in incubators – if your ducks don't go broody and you want to hatch some out, use a hen. The Hook Bill comes in three colours: plain White, Dusky Mallard and White-bibbed Dusky Mallard, which simply differs from the Dusky by having a well-defined smart white bib on the lower neck and breast.

Hook Bill drake

ORIGIN	CLASS	EGG COLOUR	NO. OF EGGS PER YEAR	
Probably Asia	Light	Blue	100–200	
	ADULT WEIGHT	**BILL COLOUR**	**LEG COLOUR**	**EYE COLOUR**
DRAKE	2.0–2.5 kg 4½–5 lb	Dusky Mallard and White-bibbed: slate grey White: pink	Dusky Mallard and White-bibbed: dark orange White: bright orange	Dusky Mallard and White-bibbed: brown White: blue
DUCK	1.6–2.0 kg 3½–4½ lb	Dusky Mallard and White-bibbed: slate grey White: pink	Dusky Mallard and White-bibbed: dark orange White: bright orange	Dusky Mallard and White-bibbed: brown White: blue

Hook Bill duck

Indian Runner

A duck that looks like a folded umbrella?

Indian Runner ducklings

Supposedly introduced into Great Britain by a ship's captain on his return from Malaysia and Indonesia. At first people thought they were penguins because of their strange upright stance and they were known as penguin ducks. They soon became popular thanks to their excellent laying capabilities, 300 eggs a year not being unknown, although the modern Indian Runner is more likely to lay between 100 and 200 – still a good yield.

Runners come in 14 recognised colours but all share the fact that they cannot fly, are easy to tame, particularly enjoy slugs and aren't as keen on swimming as other breeds, making them ideal if there is no access to a pond. The ducks are not proficient mothers, however, and if you want to breed it is best to use a broody hen who knows her job. The drakes are inclined to fight and it is not a good idea to keep more than one with a group of ducks.

Indian Runner duck

ORIGIN	CLASS	EGG COLOUR	NO. OF EGGS PER YEAR	
South Asia	Light	Blue	100–200	

	ADULT WEIGHT	BILL COLOUR	LEG COLOUR	EYE COLOUR
DRAKE	1.6–2.3 kg 3½–5½ lb	Wide variety	Wide variety	Wide variety
DUCK	1.4–2.0 kg 3–4½ lb	Wide variety	Wide variety	Wide variety

Indian Runner drakes

Magpie

An eye-catcher in black-and-white livery

Not surprisingly named for its black-and-white colouring, this breed achieved its Poultry Club Standard in 1926. It is a smart-looking duck with a black cap, back and tail, although breeders will find that it is difficult to attain a perfect specimen without spots of black in the wrong places or uneven markings. Also found in blue-and-white, chocolate-and-white and dun-and-white, in each case the colour replacing black.

This is a hardy breed that will forage for a good deal of its own food. Although classed as light it is quite a large duck and reluctant to fly, being easily contained by a low fence as long as predators are kept at bay.

Ducks' tails, drakes' tails,
Yellow feet a-quiver,
Yellow bills all out of sight
Busy in the river!
Kenneth Grahame,
Ducks' Ditty in
The Wind in the Willows

Magpie drake

ORIGIN	CLASS	EGG COLOUR	NO. OF EGGS PER YEAR	
UK	Light	Blue-green but sometimes white	180–200	
	ADULT WEIGHT	**BILL COLOUR**	**LEG COLOUR**	**EYE COLOUR**
DRAKE	2.5–3.2 kg 5½–7 lb	Yellow spotted with green	Orange with dark mottling	Dark grey
DUCK	2.0–2.7 kg 4½–6 lb	Grey-green	Orange with dark mottling	Dark grey

Magpie duck

Magpie ducklings

Muscovy

The ugly duckling in a class of its own

The Muscovy was already domesticated when Columbus arrived in America in 1492 and is native to South and Central America. It is part of a subgroup of perching ducks known as 'the greater wood ducks'. Perching is unusual in domestic ducks, most preferring to roost on the ground.

Muscovys are large, rather clumsy-looking ducks thanks to their somewhat short legs, but are nevertheless excellent swimmers and strong flyers – unless they become too heavy, in which case one wing will need to be clipped to contain them. Not the prettiest ducks, they have wart-like growths on their faces known as 'caruncles' and in some cases these can become so large as to hamper their vision. They are bred mainly for meat as it tends to be less fatty than some other breeds – you may find it labelled as Barbary Duck. These are birds with voracious appetites, and they will eat anything and everything they come across from insects to rats, mice and slugs, but also have calm temperaments and make characterful pets. Drakes can be aggressive, particularly as they get older.

Muscovy drake

Ducks make excellent mothers and can raise up to three broods a year. The incubation period is longer than most other breeds, at 35 days. They also have the advantage of quiet voices.

This breed is found in a wide range of colours: black, blue, chocolate and white as well as a range of magpie patterns, where the head has a clearly marked cap, and the back from shoulders to tail plus scapular feathers are also a solid colour.

ORIGIN	CLASS	EGG COLOUR	NO. OF EGGS PER YEAR	
The Americas	Heavy	Creamy white	30–70	
	ADULT WEIGHT	BILL COLOUR	LEG COLOUR	EYE COLOUR
DRAKE	4.5–6.3 kg 10–14 lb	Pinkish with darker shading	Pink, yellow or black	Yellow, brown or blue
DUCK	2.3–3.2 kg 5–7 lb	Pinkish with darker shading	Pink, yellow or black	Yellow, brown or blue

Muscovy duck

Orpington

A smallholder's dream in a shiny buff coat

William Cook of Kent, who bred the Orpington chicken, was the creator of this handsome duck. He crossed Indian Runners, Rouen, Cayuga and threw in a bit of Aylesbury to produce a general-purpose duck. This breed is known as 'Buff' in the USA.

The buff colour is unstable and offspring will hatch into three colours: the Buff, which is the standard where the birds should have seal-brown heads and deep buff bodies; the Blond, which is a paler version with a light grey-brown head; and the Brown, where the birds are light khaki with pencilled plumage on the females and brown head and rump on the drake.

This is an ideal backyard duck as it not only lays well, is temperamentally calm and disinclined to fly, but also produces an attractive carcass.

Orpington duck

ORIGIN	CLASS	EGG COLOUR	NO. OF EGGS PER YEAR	
UK	Light	White	150–200	

	ADULT WEIGHT	BILL COLOUR	LEG COLOUR	EYE COLOUR
DRAKE	2.2–3.4 kg 5–7½ lb	Yellow-ochre	Orange-red	Brown
DUCK	2.2–3.2 kg 5–7 lb	Orange-brown	Orange-red	Brown

Orpington drake

Pekin

A cuddly creature with a loud voice

Pekins have been found in China for over 2,000 years and arrived in the UK and USA in the 1870s. They have a rather upright carriage – although the American strain is less erect – and their fluffy feathers and rather chubby cheeks give them the look of a child's soft toy.

Their soft feathering tends to get muddy and also attracts parasites, so these ducks need plenty of water in order to keep their feathers in good condition. Their plumage is in fact cream, even yellow, in colour. They are friendly and rather noisy birds that make good pets as well as laying plenty of white eggs. They rarely go broody, don't fly and because of the thick feathering around their necks sleep stretched out flat rather than putting their heads under a wing.

This is a very popular meat breed that is often crossed with the Aylesbury to produce fast-maturing birds with good meat by 8 weeks old.

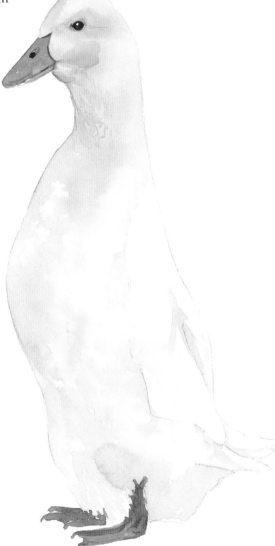

Pekin duck

Pekin duckling

ORIGIN	CLASS	EGG COLOUR	NO. OF EGGS PER YEAR		
China	Heavy	White	70–100		

	ADULT WEIGHT	BILL COLOUR	LEG COLOUR	EYE COLOUR
DRAKE	4.1 kg 9 lb	Orange	Orange	Dark lead blue
DUCK	3.6 kg 8 lb	Orange	Orange	Dark lead blue

Pekin drake

Rouen Clair

A utility bird with Mallard colouring

Similar to the Rouen and also from the Normandy area of France, the Rouen Clair differs with its slightly more upright carriage and pale ground colour of the duck known as Isabelle Clair.

This is a popular utility duck in France owing to the good-sized carcass produced and the fact that this breed is a prolific layer. The bird tends to waddle thanks to their large size and have appetites to match. Rouen Clairs are, however, easily contained as they are usually too heavy to fly.

Get your ducks in a row. To organise things well.

Rouen Clair duck with ducklings

ORIGIN	CLASS	EGG COLOUR	NO. OF EGGS PER YEAR	
France	Heavy	White	150	

	ADULT WEIGHT	BILL COLOUR	LEG COLOUR	EYE COLOUR
DRAKE	3.4–4.1 kg 7½–9 lb	Yellow with greenish tint	Orange	Brown
DUCK	2.9–3.4 kg 6½–7½ lb	Orange-ochre	Yellow-orange	Brown

Rouen Clair drake

Rouen

A beautiful 'canard' that can run to fat

Not surprisingly from the Rouen area of Normandy, this is a large duck with wild Mallard colouring. It arrived on these shores some time in the eighteenth century and was further developed until it became known as the Rouen foncé (dark Rouen) to distinguish it from its French ancestor.

Rouen ducklings

Being large, this is a duck with a voracious appetite that can become so fat it can hardly waddle. It does best on level ground with shallow water – it is mostly too heavy to leave the ground so can be contained by low fences, predators permitting.

In her 1861 *Book of Household Management* Mrs Beeton says: 'The Rouen, or Rhone duck, is a large and handsome variety, of French extraction. The plumage of the Rouen duck is somewhat sombre; its flesh is also much darker, and, though of higher flavour, not near so delicate as that of our own Aylesbury.'

Nowadays it is valued more for its size and plumage than as a commercial meat bird.

Rouen duck

ORIGIN	CLASS	EGG COLOUR	NO. OF EGGS PER YEAR	
France	Heavy	White	80–100	
	ADULT WEIGHT	BILL COLOUR	LEG COLOUR	EYE COLOUR
DRAKE	4.5–5.4 kg 10–12 lb	Green-yellow	Orange-red	Dark hazel
DUCK	4.1–5.0 kg 9–11 lb	Orange	Orange-brown	Dark hazel

Rouen drake

Saxony

A charming all-rounder and friendly to boot

As the name suggests this breed of duck was developed in Germany in the 1930s from a crossing of German Pekin, Rouen and Blue Pomeranians. Having only just become established as a breed in its own right it was almost lost during the Second World War. Rather similar in shape to the Rouen Clair but a paler shade all over, the drake has a blue-grey neck and white collar that should go all the way round its neck without a break and the duck is more of an apricot colour.

This is a true all-round bird that produces a good meaty carcass at an early age as well as laying a fair number of eggs. The ducks make caring mothers and both ducks and drakes are friendly and adaptable birds, although they can be noisy.

saxony duck and drake

saxony duckling

ORIGIN	CLASS	EGG COLOUR	NO. OF EGGS PER YEAR	
Germany	Heavy	White	100–150	
	ADULT WEIGHT	BILL COLOUR	LEG COLOUR	EYE COLOUR
DRAKE	3.6 kg 8 lb	Yellow with greenish tinge	Orange	Brown
DUCK	3.2 kg 7 lb	Yellow with brownish tinge	Orange	Brown

saxony drake

Silver Appleyard

An attractive addition to any garden

Reginald Appleyard, who created the Ixworth hen, set out to breed an all-round duck that was a good layer and also possessed a handsome carcass with the white skin that housewives preferred. This he did in the late 1930s. The Silver Appleyard also has the advantage of being early maturing and its meat not being as fatty and greasy as some duck.

Similar but paler in colouring to the Rouen Clair, the Silver Appleyard is a pretty bird and with its calm temperament is an ideal choice for the back garden. It reached the USA in the 1960s and was standardised in 1982. There is also a miniature form of this breed.

If you should rear a duck in the heart of the Sahara, no doubt it would swim if you brought it to the Nile.

Mark Twain

Silver Appleyard duck

ORIGIN	CLASS	EGG COLOUR	NO. OF EGGS PER YEAR	
UK	Heavy	White	100–150	
	ADULT WEIGHT	BILL COLOUR	LEG COLOUR	EYE COLOUR
DRAKE	3.6–4.1 kg 8–9 lb	Yellow	Orange	Dark hazel
DUCK	3.2–3.6 kg 7–8 lb	Yellow	Orange	Dark hazel

silver Appleyard drake

Welsh Harlequin

A friendly layer of lots of white eggs

The Welsh Harlequin came about as a bit of a mistake – two mutations from a flock of Khaki Campbells. These were bred on shortly after the Second World War and were originally known as Honey Campbells but were renamed Welsh Harlequins when the breeder moved to Wales.

This is an excellent laying breed with a good carcass. The ducks are reasonably efficient mothers and as a breed the birds rarely fly, are good foragers and although they enjoy swimming can manage with just a small amount of water.

They are sociable, friendly birds not known for intelligence and being pale in colour can also be more vulnerable to predator attack.

Welsh Harlequin duck

Welsh Harlequin duckling

ORIGIN	CLASS	EGG COLOUR	NO. OF EGGS PER YEAR	
UK	Light	White	150–250	

	ADULT WEIGHT	BILL COLOUR	LEG COLOUR	EYE COLOUR
DRAKE	2.3–2.5 kg 5–5½ lb	Olive green	Orange	Dark brown
DUCK	2.0–2.3 kg 4½–5 lb	Dark slate tinged with green	Dark brown	Dark brown

Welsh Harlequin drake

GEESE

Introduction

Of all domesticated birds, geese are probably the hardiest, healthiest and easiest to care for. They even make great guard dogs (though in order to do this they must use their fairly loud voices). All they require is a bit of space with grass – indeed they will not thrive without it. Ideal would be an orchard or small field, protection from predators and a few handfuls of grain.

> **Collective nouns for geese: on land – a gaggle or flock of geese; in flight – a skein, team or wedge of geese; flying close together – a plump of geese.**

All domestic geese are descended from the wild Greylag, apart from Chinese and African breeds who claim the Chinese Swan Goose as their ancestor. They have been domesticated for at least 2,000 years and along with many other things it is thought the Romans introduced them to Great Britain. All parts of a goose can be used, from the feathers to the feet – which are considered a delicacy by some. Annual goose fairs were held around Michaelmas, the Nottingham Goose Fair being the most famous in the UK.

It is a blind goose that cometh to the fox's sermon.
John Lyly

Chinese lying down

Housing and fencing

Geese are hardy birds and will quite happily live outside – but they are also vulnerable to predators so a house is a good idea. This could just be a shed, as for ducks, but they will need more space and a larger door. A solid concrete floor is the easiest to clean and can be covered with either shavings or straw. A nest box approximately 50 cm (20 in) square in the darkest corner with a couple of china eggs in a cosy hay bed might encourage your geese to lay in the house, but they are more likely to find their own spot which it will be hard to dissuade them from using.

Geese are grazers and need a grassy run – the larger the better. If they have to be contained the fence need not be high to keep them in but rather will have to cater for keeping predators out, in particular foxes. Electric netting could be used if you only have two or three birds and this has the advantage of being readily moveable.

For water requirements see page 4.

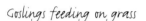
Goslings feeding on grass

Feeding

As with all domestic fowl feeding geese is not an exact science and depends entirely on the way they are kept and the amount of forage they can find for themselves. Geese with plenty of grazing will not need feeding at all during spring and summer, although a handful of grain each will help to keep them tame. If grazing is restricted and/or the birds are prolific layers or growing they will need up to 500 g (1 lb) each of mixed grain and pellets divided between morning and afternoon. Adult maintenance would be around 225 g (8 oz). They will also require a supply of grit and shell.

Breeding

The same methods described on pages 12–18 apply to geese, except that goslings will need greens of some sort as soon as they start to feed – the goose will cope with this herself but if a broody hen is used then plenty of clover, chickweed, grass and even cabbage, spinach or lettuce should be supplied and she will tear it up for the goslings to eat. If a totally artificial method is used it will be you doing this job!

The incubation period for domestic geese is 28–32 days.

Parts of a goose

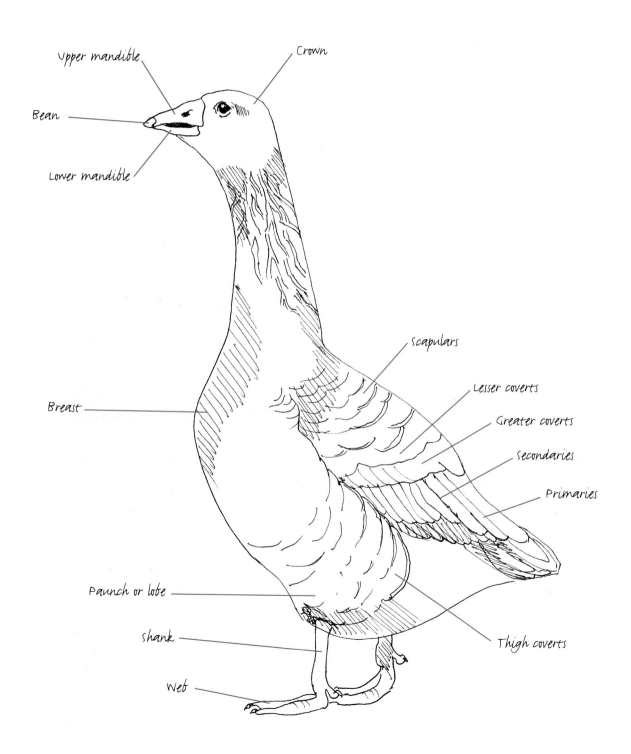

Upper mandible

Crown

Bean

Lower mandible

Scapulars

Lesser coverts

Greater coverts

Secondaries

Primaries

Breast

Paunch or lobe

Shank

Web

Thigh coverts

African

A handsome goose with a knob on its head

Along with the Chinese, the African is in fact Asiatic and evolved from the wild Swan Goose of Asia. It may also be distantly related to the Toulouse and is one of the largest and heaviest of the domestic breeds.

The African differs from the Chinese by being considerably larger and having a dewlap below its bill, although both breeds have the distinctive knob on their heads, which is larger in ganders. The knob and dewlap both grow with age. These are gentle birds that make a lot of noise, although their voices are not as loud as some. Easy to contain as they rarely, if ever, fly, they are hardy and produce a good carcass with tasty lean meat.

African geese

This breed has been recorded in the UK since the seventeenth century, although it did not achieve its Breed Standard until 1982 – it was standardised in the USA in 1874. Apart from the Brown or Grey, which have black knobs and dark orange legs, Africans are also found in Buff, with pinkish-brown knob and light orange legs, and White, where the knob and legs are orange.

A tailor's smoothing-iron is called a goose as its handle resembles the neck of a goose.

Brewer's Dictionary of Phrase and Fable

ORIGIN	CLASS	EGG COLOUR	NO. OF EGGS PER YEAR	
China	Heavy	White	10–40	

	ADULT WEIGHT	BILL COLOUR	LEG COLOUR	EYE COLOUR
GANDER	10.0–12.7 kg 22–28 lb	Bill and knob black White: orange	Brownish-orange White: orange-yellow	Dark brown White: blue
GOOSE	8.2–10.9 kg 18–24 lb	Bill and knob black White: orange	Brownish-orange White: orange-yellow	Dark brown White: blue

African goose

African gosling

American Buff

Caring parents in apricot coats

The largest of the medium-sized geese, the American Buff is similar in plumage design to the Toulouse and Brecon Buff. Its colouring is sometimes referred to as 'apricot-fawn' as it is slightly paler orange and less pink than other Buffs, although the bird also has dual lobes. The coloured feathers are all edged with pale cream but the lower parts are pure white, contrasting with bright orange legs and webs – if the legs look pale this is a sign either that the goose is laying or that there is a dearth of green grass.

American Buffs are excellent caring parents, calm and docile, and would make a good choice for the smallholder as they also produce a good-sized carcass. Ganders can be mated with three to five geese.

American Buff goose

ORIGIN	CLASS	EGG COLOUR	NO. OF EGGS PER YEAR	
USA	Heavy	White	15–25	

	ADULT WEIGHT	BILL COLOUR	LEG COLOUR	EYE COLOUR
GANDER	10.0–12.7 kg 22–28 lb	Orange with pale pink nail	Orange	Dark hazel with orange cilium
GOOSE	9.1–11.8 kg 20–26 lb	Orange with pale pink nail	Orange	Dark hazel with orange cilium

American Buff gander

Brecon Buff

An escapologist from the hills

Developed in Wales in the 1930s by crossing Buff 'sports' with Grey and White geese, this breed is slightly smaller than the American Buff, in fact it is one of the smallest breeds of goose. It is a hardy, active breed that does well in the hilly damp climate of its origin.

Brecons make good mothers but can be clumsy with their eggs and some people consider it safer to use an incubator and then reintroduce the goslings – but Brecons are wily birds and will only accept Buff goslings, they won't be fooled into raising a different breed.

With their pink bills and pink legs, dark brown eyes and feathers edged with cream, a flock of Buffs would grace any farmyard. And although they rarely fly they are adept at escaping.

Brecon Buff goose

ORIGIN	CLASS	EGG COLOUR	NO. OF EGGS PER YEAR	
UK	Medium	White	10–30	

	ADULT WEIGHT	BILL COLOUR	LEG COLOUR	EYE COLOUR
GANDER	7.3–9.1 kg 16–20 lb	Pink	Pink	Dark brown with pink cilium
GOOSE	6.3–8.2 kg 14–18 lb	Pink	Pink	Dark brown with pink cilium

Brecon Buff gander

Buff Back

An all-rounder with saddleback plumage

Sometimes called 'saddleback' or 'pied' the Buff Back differs from the similarly coloured Pomeranian by the fact that it has a dual-lobed paunch and bright orange bill, legs and webs, whereas the former's bill is pinker and legs and webs red-orange with a single-lobed paunch. Their colour pattern is the same as the Grey Back and they only differ in the shade of their feathers.

The Buff Back is typical of the geese found around the Baltic and have been known in the area for centuries. A good all-rounder that rarely flies, Buff Backs can be kept happily as trios and will sit and rear their young as long as they have access to good-quality grass.

Buff Back gander

ORIGIN	CLASS	EGG COLOUR	NO. OF EGGS PER YEAR	
Europe	Medium	White	25–35	
	ADULT WEIGHT	BILL COLOUR	LEG COLOUR	EYE COLOUR
GANDER	8.2–10.0 kg 18–22 lb	Orange	Orange	Blue
GOOSE	7.3–9.1 kg 16–20 lb	Orange	Orange	Blue

Buff Back goose

Buff Back gosling

Chinese

Asian chic with a loud voice

Elegant, active and distinctly talkative, the Chinese is closely related and descended from the wild Swan Goose of Asia. Similar to the African, with its long neck and colouring, the Chinese is considerably smaller. Both sexes have a large black knob on their heads, larger in the male, although the knob is orange in White birds.

George Washington kept a flock of these geese at his Mount Vernon Farm and they are also famous for being used as watchdogs by the Ballantine whisky distillery in Dunbartonshire, thanks to their readiness to use their voices. Ganders can be ferocious if goslings are around and are quite capable of seeing off dogs and even foxes – a fact that should be taken into account if children are present. They are intelligent birds and can take a dislike to a particular person or breed of dog.

These geese are prolific layers and characterful birds that make good pets, being easy to tame and they very rarely fly.

Chinese White goose

Chinese gosling

ORIGIN	CLASS	EGG COLOUR	NO. OF EGGS PER YEAR	
China	Light	White	40–70	
	ADULT WEIGHT	BILL COLOUR	LEG COLOUR	EYE COLOUR
GANDER	4.5–5.4 kg 10–12 lb	Black	Dark orange White: orange-yellow	Dark brown White: blue
GOOSE	3.6–4.5 kg 8–10 lb	Black	Dark orange White: orange-yellow	Dark brown White: blue

chinese goose

Czech

An active bird with a temper

A small, finely-boned goose from Bohemia, the Czech is quite a good layer that starts its breeding season early in the year. They are active, alert birds that do have the bad reputation of being aggressive, although this is mainly restricted to when goslings are around or the goose is sitting.

They have been bred in the UK since the 1990s, although the strain that is now found in Germany has changed its name to Bohemian and is slightly shorter in the body and longer in the leg than those found in the Czech Republic. This breed is only found with totally white plumage.

"All his swans are turned to geese."
All expectations end in nothing.
Brewer's Dictionary of Phrase and Fable

Czech goose on water

ORIGIN	CLASS	EGG COLOUR	NO. OF EGGS PER YEAR	
Czech Republic	Light	White	10–25	

	ADULT WEIGHT	BILL COLOUR	LEG COLOUR	EYE COLOUR
GANDER	5.0–5.5 kg 11–12 lb	Orange	Orange-red	Blue with orange cilium
GOOSE	4.0–4.5 kg 8½–10 lb	Orange	Orange-red	Blue with orange cilium

Czech gander

Embden

A striking bird with a gentle character

This is a very old breed from northern Germany known since the thirteenth century when the town of Emden was spelt with an additional 'b'. It was originally known in the USA as the Bremen and has been standardised since 1874. One of the largest breeds of goose, the Embden is a tall, upright, imposing-looking bird that has a gentle character – apart from during the breeding season.

The birds start laying early in the season but their large size makes them rather clumsy parents and they can easily break their own eggs or squash the goslings as they hatch. Incubating and hand rearing is recommended, as a result of which the birds become very tame.

Embdens are only found in pure white although the goslings differ up until 3 weeks old – the males being a lighter shade of grey.

Embden preening

ORIGIN	CLASS	EGG COLOUR	NO. OF EGGS PER YEAR	
Northern Europe	Heavy	White	10–30	
	ADULT WEIGHT	BILL COLOUR	LEG COLOUR	EYE COLOUR
GANDER	12.7–15.4 kg 28–34 lb	Orange	Orange	Light blue
GOOSE	10.9–12.7 kg 24–28 lb	Orange	Orange	Light blue

Embden goose

Embden gosling

Grey Back

The perfect backyard goose

This breed is one of several that originated in Northern Europe, known as 'saddlebacks' thanks to their pied plumage pattern. The Buff Back only differs in colour but the Pomeranian, which is also a pied goose, has a single lobe compared to the former's dual-lobed paunch.

According to Chris and Mike Ashton of Ashton Waterfowl: "The buff dilution is sex-linked. A Buff Back male bred to a Grey Back female will produce only Buff Back female goslings. Its male goslings will look like Grey Backs, but a proportion of their offspring will show buff plumage."

This is a strain that can be reproduced – Grey geese crossed with White geese can produce Grey Backs and Buff ganders crossed with White geese produce Buff Backs in two generations.

However they are produced this is an excellent medium-weight bird that produces a good plump carcass, rarely flies and tends to go broody, making good parents – a perfect backyarder.

Grey Back goose

ORIGIN	CLASS	EGG COLOUR	NO. OF EGGS PER YEAR	
Europe	Medium	White	30–40	
	ADULT WEIGHT	BILL COLOUR	LEG COLOUR	EYE COLOUR
GANDER	8.2–10.0 kg 18–22 lb	Orange	Orange	Blue
GOOSE	7.3–9.1 kg 16–20 lb	Orange	Orange	Blue

Grey Back gander on water

Pilgrim

A medium-sized goose with a pleasant disposition

There is some dispute over how the Pilgrim got its name but what is known is that it was nothing to do with the Pilgrim Fathers and the breed was not on board the Mayflower. Possibly they were named by Oscar Grow, a well-known waterfowl breeder in the early twentieth century who claims he named them after his family's pilgrimage to Missouri during the Great Depression.

The Pilgrim is known as an autosexing breed because the males and females are different colours. The gander is white, although a little grey may creep in here and there, and the goose is mostly grey.

Pilgrims have a reputation as calm and charming birds. They are good foragers, make excellent parents and produce a good plump carcass with a fast growth rate. What more could anyone want?

What is sauce for the goose is sauce for the gander.
Traditional saying

Pilgrim gander

ORIGIN	CLASS	EGG COLOUR	NO. OF EGGS PER YEAR	
Europe	Medium	White	20–40	
	ADULT WEIGHT	BILL COLOUR	LEG COLOUR	EYE COLOUR
GANDER	6.3–8.2 kg 14–18 lb	Orange	Orange	Blue-grey
GOOSE	5.4–7.3 kg 12–16 lb	Orange	Orange	Hazel

Pilgrim goose

Pomeranian

A loud-mouthed northern breed

Pomerania is a former province on the southern coast of the Baltic and these geese have been known there since at least the sixteenth century. Found in three colourways, Grey Saddleback, Grey and White, it is the Saddleback that is the standardised colour in the UK. Pomeranians were taken to America with early German settlers but did not achieve an American Poultry Association standard until 1977. The single lobe or paunch is a distinguishing feature.

As with all birds different strains can produce different characteristics but as a whole this is a calm and friendly breed, although ganders will nearly always become aggressive during the breeding season. They are lively birds and like to use their voices, a fact that should be taken into account if there are neighbours nearby – but this also means they make excellent guard birds. Thanks to its northern origins this is a very hardy breed.

Gossamer: a fine, filmy substance consisting of cobwebs spun by small spiders. The word is supposedly derived from goose summer and refers to goose down which is also light and delicate.

Pomeranian

ORIGIN	CLASS	EGG COLOUR	NO. OF EGGS PER YEAR	
Europe	Medium	White	30–75	
	ADULT WEIGHT	BILL COLOUR	LEG COLOUR	EYE COLOUR
GANDER	8.2–10.9 kg 18–24 lb	Pink to orange-red	Orange-red	Blue or brown with red cilium
GOOSE	7.3–9.1 kg 16–20 lb	Pink to orange-red	Orange-red	Blue or brown with red cilium

Pomeranian

Roman

A graceful possessor of a raucous voice

This ancient breed of little goose is credited with alerting the Romans with their alarm calls when the Gauls attacked during the night in 365 BC. These birds enjoy using their voices – a fact that should still be taken into account today.

The Roman is an elegant plump bird that can become very tame, although as with many breeds traits can go with the strain. As a rule, however, this is a placid good-natured breed ideal for smallholders, although the very fact that they are smaller than most breeds also means that they are capable of flight so one wing will need to be clipped, certainly until they know their home.

Crested Roman

Romans also come with a crested variety – this is the type most often found in North America. The crest is a simple tuft of feathers that should be right on the top of the head beginning just behind the eyes. Both types make good mothers and lay a reasonable amount of eggs.

Roman gosling

GEESE SAVE THE CAPITOL
When the Gauls invaded Rome a detachment in single file clambered up the hill of the Capitol so silently that the foremost man reached the top without being challenged; but while he was striding over the rampart, some sacred geese, disturbed by the noise began to cackle and awoke the garrison. Marcus Manlius rushed to the wall and hurled the fellow over the precipice. To commemorate this event the Romans carried a golden goose in procession to the Capitol every year (390 BC).

Brewer's Dictionary of Phrase and Fable

ORIGIN	CLASS	EGG COLOUR	NO. OF EGGS PER YEAR	
Europe	Light	White	40–60	
	ADULT WEIGHT	BILL COLOUR	LEG COLOUR	EYE COLOUR
GANDER	5.4–6.3 kg 12–14 lb	Orange-pink	Orange-pink	Light blue
GOOSE	4.5–5.4 kg 10–12 lb	Orange-pink	Orange-pink	Light blue

Roman gander

Sebastopol

Has it been through a hedge backwards?

Sebastopols are rather extraordinary-looking geese because their feathers are much longer than the average goose and twist and turn in all directions giving them their characteristic dishevelled look. Lord Dufferin is credited with introducing this breed from the Black Sea when he returned from the Crimean War and they are still common around the Danube, and sometimes even known as the Danubian.

Not all Sebastopols have curly feathers. There is also a smooth-feathered variety, although this has long trailing feathers from scapulars and thighs that can drag along the ground. The smooth-feathered variety can fly, unlike the curly variety where even the flight feathers are loose and tangled, making flight impossible.

Certainly a conversation piece, Sebastopols have the reputation of being charming, hardy birds that would grace any garden.

sebastopol goose

ORIGIN	CLASS	EGG COLOUR	NO. OF EGGS PER YEAR	
Europe	Light	White	30–50	
	ADULT WEIGHT	BILL COLOUR	LEG COLOUR	EYE COLOUR
GANDER	5.4–6.3 kg 12–14 lb	Orange-pink	Orange-pink	Bright blue
GOOSE	4.5–5.4 kg 10–12 lb	Orange-pink	Orange-pink	Bright blue

sebastopol goose

sebastopol gosling

Shetland

A tough bird from northern climes

The Shetland Islands off the north coast of Scotland are remote islands boasting several unique breeds such as sheep and ponies – the Shetland goose is another. Harsh climates are inclined to produce small tough creatures and this goose is no different, being hardy and able to forage for a good deal of its own food.

This is an autosexing breed, the gander having white plumage and the goose a grey saddleback and mottled head and neck, and the lower body is white. The goose will lay a clutch of up to 20 eggs and the pair make very good, protective parents.

shetland goose

ORIGIN	CLASS	EGG COLOUR	NO. OF EGGS PER YEAR	
UK	Light	White	20–30	
	ADULT WEIGHT	**BILL COLOUR**	**LEG COLOUR**	**EYE COLOUR**
GANDER	5.4–6.3 kg 12–14 lb	Pink	Pink	Blue
GOOSE	4.5–5.4 kg 10–12 lb	Pink	Pink	Blue

shetland gander

Skåne

Specially designed for a goose supper

Known as Skånegås in Sweden this goose is descended from those supposedly brought back by Swedish soldiers in the eighteenth century. They were created from a mixture of breeds but with a strong proportion of Pomeranian. Fast-maturing birds, they reach a good weight in six months – in time for the Swedish National Goose Day on 11 November (St Martin's Day). Traditionally goose was eaten for dinner on Goose Day Eve.

These birds are very easy to fatten, producing high-quality meat thanks to their white skin and pale flesh. They generally have lovely temperaments although they tend not to lay many eggs and go broody easily.

A goose quill is more dangerous than a lion's claw.
English proverb

skåne goose

ORIGIN	CLASS	EGG COLOUR	NO. OF EGGS PER YEAR	
Swedish	Heavy	White	20–30	
	ADULT WEIGHT	BILL COLOUR	LEG COLOUR	EYE COLOUR
GANDER	11–14 kg 24–31 lb	Orange-yellow	Pinkish-orange	Blue-grey
GOOSE	9–11 kg 20–24 lb	Orange-yellow	Pinkish-orange	Blue-grey

skåne goose

Steinbacher

A fighter with a placid streak

Originally bred in Eastern Europe as fighting geese for a sport similar to cockfighting, this breed is actually no more aggressive than any other. In fact it is easy to tame into characterful friendly pets. It does have a determined streak, however, and in the breeding season males will fight almost to the death.

Steinbachers are attractive geese despite their slightly Roman-nosed bills, with an upright proud stance and are found with blue, lavender, grey, cream and buff plumage. Egg production depends on the strain – some laying up to 40 eggs a year and others considerably fewer.

A wild goose never reared a tame gosling.
Irish proverb

steinbacher goose

ORIGIN	CLASS	EGG COLOUR	NO. OF EGGS PER YEAR	
Germany	Light	White	20–40	
	ADULT WEIGHT	BILL COLOUR	LEG COLOUR	EYE COLOUR
GANDER	6.0–7.0 kg 13–15 lb	Orange with black bean	Bright orange	Dark brown with orange cilium
GOOSE	5.0–6.0 kg 11–13 lb	Orange with black bean	Bright orange	Dark brown with orange cilium

steinbacher goose

Toulouse

A French favourite for the table

Specifically developed as table birds by the French, this is the breed used to produce foie gras. It was brought to the UK by the Earl of Derby in the 1840s and also appeared in the USA at about this time. The bird found in the UK is a much heavier type than the French, with a large dewlap and keel that almost touches the ground.

In France these geese are kept in large numbers in open pasture. They tend not to fly so need only a normal-height fence to contain them, although predators must be kept out. They are calm birds that are good foragers and although hardy will do better with shelter from rain. Some breeds are particularly prone to 'angel wing', where the flight feathers become twisted.

Toulouse goose

ORIGIN	CLASS	EGG COLOUR	NO. OF EGGS PER YEAR	
France	Heavy	White	20–40	
	ADULT WEIGHT	BILL COLOUR	LEG COLOUR	EYE COLOUR
GANDER	11.8–13.6 kg 26–30 lb	Orange	Orange Buff: grey	Dark brown with orange cilium White: Blue
GOOSE	9.1–10.9 kg 20–24 lb	Orange	Orange Buff: grey	Dark brown with orange cilium White: Blue

Toulouse goose

West of England

An old breed with a useful accomplishment

An autosexing breed where the gander is white and the goose develops a grey saddleback and neck, this is one of the few breeds where the goslings can be sexed when they hatch. This is an old breed known in the UK since before 1600 – it may well have featured in the ancestry of the Pilgrim goose.

These are friendly birds that are very easy to manage and are good foragers. They make great backyard birds. The gander, as with most breeds, can become aggressive while the goose is sitting but will be calm and sociable for most of the year. Traditionally the goose will lay from St Valentine's Day until mid-April and they make excellent parents. Although hardy these geese will need shelter from the elements and protection from predators.

Wild goose chase. A hopeless quest. This old phrase was first recorded in Shakespeare's *Romeo and Juliet* in 1597.

West of England goose

ORIGIN	CLASS	EGG COLOUR	NO. OF EGGS PER YEAR	
UK	Medium	White	20–50	

	ADULT WEIGHT	BILL COLOUR	LEG COLOUR	EYE COLOUR
GANDER	7.3–9.1 kg 16–20 lb	Orange	Orange or pink	Blue
GOOSE	6.3–8.2 kg 14–18 lb	Orange	Orange or pink	Blue

West of England gander

GUINEAFOWL

Introduction

Guineafowl have been around and domesticated for thousands of years – a guineafowl even appears as an ancient Egyptian hieroglyph of the sound NH (pronounced 'narh') dating to 2,500 BC.

It is the Romans who are credited with introducing guineafowl to Britain, but they are thought to have died out when the Romans left and weren't documented again until the fourteenth or fifteenth centuries. Their name comes from Guinea on the West African coast and is a Tuareg word 'aginau' meaning 'black people'. Young guineafowl are known as keets and this is supposedly from the old Nordic word 'keetling'.

In Greek mythology the hero Meleager (whose name means guineafowl) was slain after defending the honour of the huntress Atalanta. The goddess Artemis turned his sisters Gorge and Deianira (the wife of Heracles) into guineafowl, which Artemis considered her sacred birds. However, the god Dionysus begged Artemis to return the two women (known as the Meleagrids) to their human form, and she did.

Guineafowl keets

Guineafowl are independent creatures that are wild at heart. It is no good buying adult or even young birds and expecting them to remain on your property – they will simply fly away unless totally caged. The best way to start with guineas is to hatch your own with a broody hen or incubator or buy day-olds if you can find any. Even home-reared birds will start to roost up trees as soon as they are given the chance. These birds are very difficult to sex, and the easiest way is by listening – the female bird has a two-syllable call described as 'buck wheat, buck wheat' and the male a single-syllable screech. This screech can be heard for a great distance and is reminiscent of a machine gun – this may not enamour your neighbours when your guineas spot a fox passing by at 4 am in the summer and make their presence known.

Guinea eggs have an exceptionally tough shell and can travel by post successfully, so it is possible to hatch your own either by using an incubator or a broody hen. Incubation time is long – 28 days as opposed to 21 in chickens – so the broody needs to be a steady and patient creature. Female guineas will lay and go broody but they do so on the ground, so if they are not enclosed are exceptionally vulnerable to fox and other predator attacks. The female will create a small depression where she will lay her eggs – other females may come and lay theirs in the same place until there are as many as 30 eggs, at which point one will go broody and incubate them. If at all possible try and find out where they are laying, but this will not be easy. If they think you have seen their private spot or are taking their eggs they will change it, so leave a few china or marked eggs in the nest. A sitting guineafowl free range will almost certainly be found by a fox. Birds tend not to go broody if they are kept enclosed but will still lay.

Being insect-eaters guineas do less scratching than hens but beware if you keep bees – they will stand outside the hives and literally snatch the bees as they come out.

Guineafowl come in a variety of colours, the most common of which are:

- Pearl – the original 'wild' colour with white spots
- Pearl and White (sometimes called pied) – wild colour with white underneath
- Lavender – a pale grey with white spots
- Lavender Pied – pale grey / white spots and white underside
- Mulberry – a dark purple body with buff head
- Mulberry Pied – dark purple body with white underside
- Buff – buff / khaki colour
- Chocolate
- White

Pearl guineafowl

Mulberry guineafowl

White guineafowl

Housing and fencing

Guineafowl are happiest free ranging but as mentioned before they will not stay around unless they have either hatched or come as day-olds to your property. They will roost high up in trees in all weathers and once they have had a taste of freedom it will be impossible to lure them back into a shed or barn.

Lavender guineafowl

Therefore if you want them contained they will have to be in a very large enclosure with netting over the top as well – as an example if you have 30 to 40 birds you would need an enclosure the size of a tennis court. The sides will have to be predator proof, as described on page 22.

Guineafowl feathers

Pearl guineafowl

Feeding

Guineafowl can be fed normal hen food once they are 10 to 12 weeks old. If totally confined a small handful of pellets in the morning and one of mixed corn in the evening will be fine. Keets should have a high-protein diet – turkey or game bird starter crumbs would be best but if these are unobtainable try and find unmedicated crumbs.

If your birds are free range they will get most of their food for themselves but to keep them as tame as possible and at least on your property, give them mixed corn – twice a day if they will come to get it.

There should always be plenty of clean water available – a normal poultry tower would be ideal.

Water tower

Collective noun for guineafowl:
a rasp of guineafowl.

Breeding

Guineafowl make excellent mothers as long as they have laid their clutch out of the range of foxes and other predators. The eggs can also be successfully put to a broody hen, although the incubation time is 28 days so she will have to be a tight-sitting bird, or in an incubator. Thanks to the exceptionally thick shells, candling will be extremely difficult and it will probably be best just to wait out the time and hope to see them pipping at the end of the month (see page 14).

Parts of a guineafowl

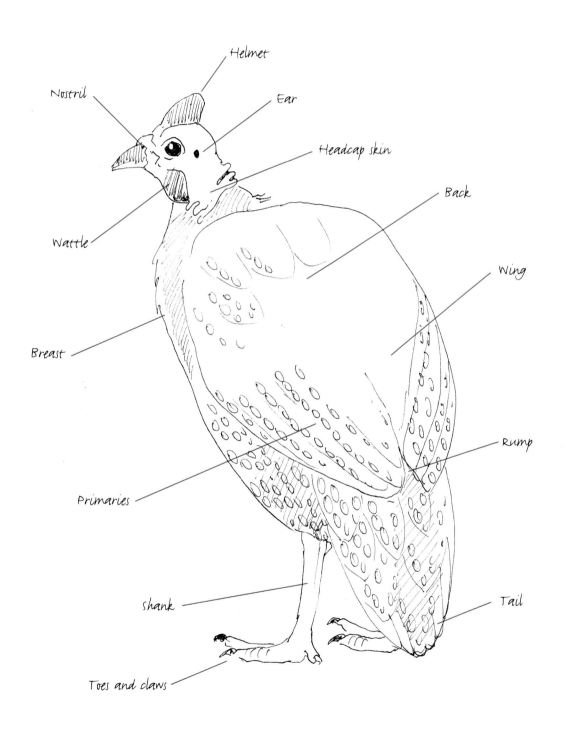

Helmet

Nostril

Ear

Headcap skin

Back

Wattle

Wing

Breast

Rump

Primaries

Shank

Tail

Toes and claws

PEAFOWL

Introduction

The peacock is the national bird of India and features in Hindu mythology. Vishnu always wears a peacock feather in his headband and the bird is considered sacred by Indians. While the peacock is revered in India and considered good luck, in the West it is thought unlucky to bring peacock feathers into the house because of the 'evil eye' in the tail feather.

In fact the train is not actually the peacock's tail but is made up of 100–150 upper tail coverts that are supported by 20 retrices or true tail feathers. Male birds won't develop a full train until they are 3 years old. They display during the breeding season to attract a female. The peahen will start to lay in her second year and lays between 3 and 12 eggs in a clutch – if left to her own devices she doesn't build a nest but simply finds a place hidden in bushes.

Before you even consider keeping peacocks you must be aware of the noise they make. Their cry carries some considerable distance and during the breeding season, which lasts about 4 months, they will shriek constantly – and this won't be confined to daylight hours. The rest of the year they are quieter but still give their mournful cry every now and then.

Peacock feather

If a peacock cries more than usual, or out of its time, it foretells the death of someone in the family to whom it doth belong.

Paracelsus

Housing and fencing

Peafowl can be kept in an enclosure or aviary but this will necessarily have to be large. They are good flyers and from choice will roost up trees. If you are considering buying in a pair of fully grown peafowl then you will have to keep them contained for several weeks or they will simply fly away. A better way would be to start with eggs, day-olds or young poults and keep them contained, ideally in a barn with rafters that they can roost in – this way they will know that this is home and when you finally let them out, they should return every night on their own. If you can manage to shut them up at night this will not only solve the problem of 3 am shrieks it will protect them from predators. Although safe up a tree during the night they will fly down at 5 am or so in the summer months when it gets light and this is when the fox will be waiting.

In Feng Shui, peacocks are a symbol of beauty and fidelity. Displaying the image of a peacock in the southern region of a bedroom is said to encourage a lasting and loving partnership.

These are hardy birds and will be fine roosting up trees all year round but if you want them to use a house, or have them in an enclosure with a house, make sure it is large enough – at least 3 m by 3 m (10 foot by 10 foot) with perches that they can flap up to and swoop down from without injuring themselves. Use shavings or straw on the floor that can be easily cleaned out on a regular basis.

They will appreciate a dust bath, and this is important for feather maintenance, so if contained they will need a dry covered area. If free range they will find their own spot – probably in a flower bed!

Peafowl drink a lot of water – small buckets or towers used for chickens are fine.

Indian peahen

Indian peacock

Feeding

In a natural environment peafowl eat seeds, insects and vegetation but even if your birds are totally free range they will still do better (and of course remain tame) if you feed them once a day. Give them game bird or turkey maintenance pellets – a handful each.

Feeding peafowl inside a building makes catching them a likelier possibility, as trying to catch a fully grown free-range bird will be almost impossible unless it is exceptionally tame (see page 120).

If enclosed they will need feeding twice a day – the pellets in the morning and mixed corn in the afternoon. Day-old chicks should be fed game bird or turkey crumbs ad lib until they are around 3 or 4 weeks old when they can move on to growers pellets. Start adding the mixed corn when they are about 12 weeks old. From week 18 or so they can be fed the maintenance pellets. How much you give will depend entirely on how much they can forage for themselves, but if you give the pellets in containers and there are any left after 20 minutes then this is probably too much. You don't want any food left lying around as this will attract rats. They will enjoy leftovers from your kitchen such as boiled potatoes, pasta, rice or vegetables.

They will also require mixed grit and oyster shell.

Collective nouns for peafowl: a muster, ostentation or pride of peafowl.

White peachicks

White peacock

Breeding

Ideally the peahen should be at least 2 years old and the peacock 3 years old before you consider hatching eggs. One peacock will keep up to five hens fertile. The hen generally lays around 5–9 eggs, but sometimes more, and will be happy to incubate and rear her chicks herself. If she is free range she will find her own spot and you probably won't see her once she is sitting until she reappears with a line of peachicks behind her. She is vulnerable to foxes though at this time. Peahens are excellent mothers and will look after their broods for up to 6 months. The incubation for peachicks is 28 days.

A broody hen will do the job for you as well but peachicks imprint themselves on their mothers, so the hen will either have to stay with them or be very carefully weaned off. She may well lose interest while they still require warmth, particularly at night, so a lamp may be needed. Exactly the same method and care of the broody applies as described on page 12.

An incubator can also be used – candle after 10 days and 20 days, but otherwise follow the method on page 15.

How to catch a peacock

There may come a time when you want to catch one of your peacocks and if they are free range this will be a problem. One way is to try and drive the bird into a shed or barn and corner it, grabbing it by the legs – never the feathers, which may pull out. Hold the legs firmly in one hand and support the body with the other.

Another way is to use a very large fishing net and drop it over the bird's head and breast before grabbing the legs. Whatever method you use, peacocks have very sharp beaks and will not take kindly to being manhandled.

Parts of a peacock

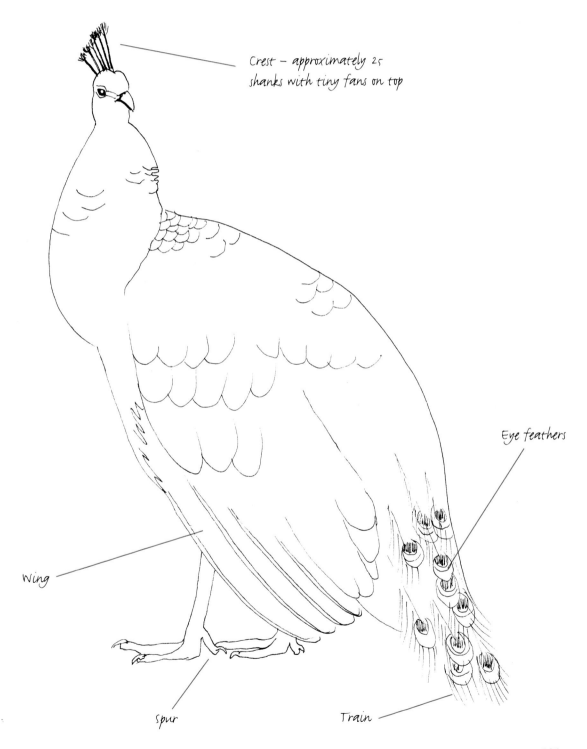

Crest — approximately 25 shanks with tiny fans on top

Eye feathers

Wing

Spur

Train

QUAIL

Introduction

Quail have made up part of the diet of humans since Stone Age times or before, and also appear as an Egyptian hieroglyphic. Classed as a game bird, there are over 40 different species of quail throughout the world, including Bobwhite and Californian, but only Coturnix have become truly domesticated and these come in a wide variety of colours.

Quail are Galliformes and belong to the Phasiandae family, which also includes pheasants and partridges. Like them they spend most of their time on the ground and only burst into the air as a last resort. This behaviour is a problem if the birds are kept enclosed, because if frightened they will 'boink' or spring up and hit their heads on the roof of the pen or shed. A lower covering of soft netting underneath the roof netting might save them from injury.

> Collective nouns for quail:
> a covey or bevy of quail.

Even these tiny birds have voices, the males being louder than the females, although not as loud as some other forms of poultry. If kept in small cages then the males should be separated from each other but if their run is spacious they can cohabit quite happily.

The parrot utters one cry,
the quail another.

Latin proverb

Tuxedo quail

Tuxedo quail

scarlet Tuxedo quail

Quail eggs

A quail egg provides around 14 kcal, 1 g of fat and 1.2 g of protein.

Housing and fencing

Quail can live in a fairly small area if only a few are kept. A converted rabbit hutch or other cage might house 10 or 12 and the best kind of bedding is wood shavings, although some commercial breeders use a kind of weld mesh that the droppings can fall through for ease of cleaning, although this is not as comfortable for the birds. Purpose-built aviaries are also available. Birds will be happier if kept in a more natural environment of house and small covered run. This must be predator-proof – a rat or even a sparrowhawk will take a quail – and as mentioned earlier some form of overhead protection for the birds will save them from injury. Make your run as natural as possible, with branches and places to hide. Females can lay up to 200 eggs a year but may not take much notice of where – a cosy nest box on the floor might tempt them but most likely they will just lay where they happen to be at the time.

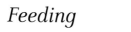

Harlequin male quail

Feeding

Quail can be fed normal poultry feed – crumbs to start with, although those formulated for turkeys or game birds have a higher protein level than those for chickens, which helps speed growth. Quail layers pellets are also available, but tend to be expensive and you would have to order them specifically from your feed merchant. Clean water must always be available – small tower units are fine or specially designed cage bird or rabbit drinkers can be used.

Japanese female quail

Japanese male quail

Japanese female quail

Breeding

Coturnix quail can be kept in pairs or trios or one male will fertilise five or six females. The hen birds will start laying at around 5 or 6 weeks of age. As soon as the females are fertile the male will start to produce foam balls – a bit like shaving foam – a urinary product thought to indicate fertility. In captivity the birds tend not to go broody so either you will have to enlist the help of a very small and careful broody bantam or use an incubator. The latter is probably the most reliable as the eggs and chicks are so small and fragile. Coturnix varieties hatch at 18 days and candling is possible, but fiddly, and this should be done at around day 6 or 7, although as the incubation time is so short it hardly seems worth risking disturbing the eggs. Try to find perfect eggs and don't keep them for more than 7 days before starting incubation. Follow the method described on page 15. Once hatched and dry transfer the chicks to a small enclosure – a cardboard box would be fine to start with – but watch carefully at first that your lamp is neither too hot nor cold. The temperature should be 35°C (95°F) at first but the heat should be gradually decreased to no heating after about 3 weeks, depending on the time of year and outside temperature. Quail are ace escapees so make sure you have a cage to move them to with no gaps or cracks and high enough sides – they can jump surprisingly high. Feed chick crumbs ad lib and supply water in containers, being careful that the tiny chicks cannot drown. Placing gravel in the trough might be the answer.

Immature Japanese quail

Children of the Ritz, sleek and civilized ...
We know just how we want our quails done,
and then we go and have our nails done.

Noel Coward

Quail chick

With practice you will be able to sex your birds from about 3 weeks onwards. The male will begin to develop a more reddish-brown chest and the female spots on hers. Other colour varieties are not as easy to differentiate although eventually the female will be larger. The male has a shrill tri-syllable call, much louder than the female's voice. Finally the male will begin to produce foam balls – if you gently squeeze the cloaca gland foam will squirt out. In the wild these foam balls are territory markers and the quail is the only bird that produces them.

Parts of a quail

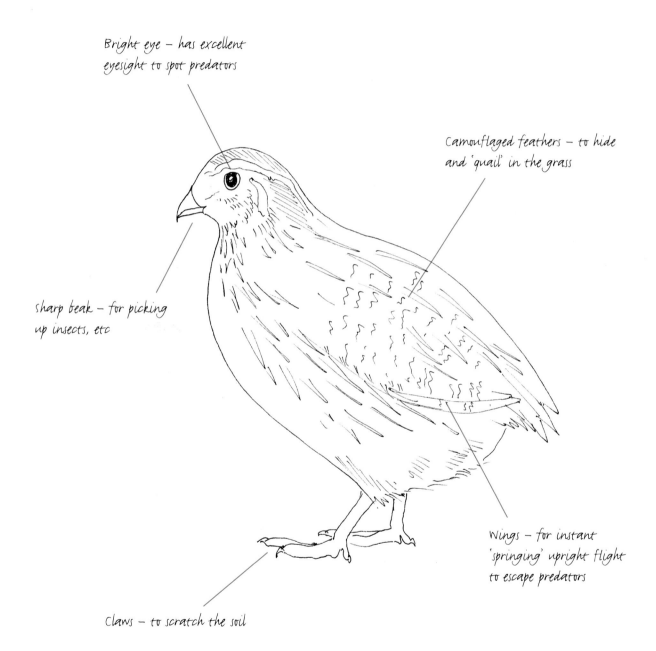

Bright eye – has excellent eyesight to spot predators

Camouflaged feathers – to hide and 'quail' in the grass

sharp beak – for picking up insects, etc

Wings – for instant 'springing' upright flight to escape predators

Claws – to scratch the soil

TURKEYS

Introduction

Turkeys originate from North and Central America and were first domesticated by the Aztec Indians. They were an important part of their society, being reared not only for their meat but also their feathers, which were used for ceremonial purposes. There are many theories as to how the turkey got its name but no one really knows for sure. One theory is that the bird looked like a Turkish soldier or turkoman, while another suggests that the Tamil word in India for peacock, 'tali', was the source. Yet another is that Luis de Torres, a doctor who travelled with Christopher Columbus, called the bird 'tukki' or 'big bird' in Hebrew.

Turkeys first arrived in Europe around 1500 – King Ferdinand of Spain requested that all ships returning from the Caribbean should bring some back and they became exotic curiosities as well as meat for the very rich. From Spain they eventually came to England and were farmed in East Anglia where the Norfolk Black was created. This in turn was taken back to the USA with the Pilgrim Fathers and became associated with Thanksgiving Day. It is still the mainstay of the Christmas table in the UK, however.

Turkeys are charming, rather gentle birds that are easy to tame and can live to 9 years or more. They have higher blood pressure than other domestic fowl and can suffer heart attacks if chased or otherwise upset. They are vociferous, which should be taken into account if you have neighbours – a stag's gobble can be heard over a mile away.

I have no desire to crow over anybody or to see anybody eating crow, figuratively or otherwise. We should all get together and make a country in which everybody can eat turkey whenever he pleases.

Harry S. Truman

Bronze turkey

The stag or male bird has fleshy bulbous protuberances on its head and neck called caruncles – these can change colour from red through blue to pink, depending on the bird's mood – and also a snood, which is the fleshy appendage that hangs over the beak. The snood extends when the bird displays but can be contracted when not displaying. The hen also has a small snood but hers cannot extend. Turkeys have magnificent tail feathers, usually 18 that fan out during display. They also drag their wings on the ground while strutting. They are happy to display all day if there is a hen around, or in fact anything that they want to impress. They fan their tail as well as drum, which is a sort of rapid stamping on the spot.

From this to this in under one minute

Turkeyspeak

Adult stags tend to gobble a good deal of the time and also make a sort of 'pooumf' or deep 'donk' noise when displaying, which is a displacement of air in the region of the crop. They also make a quiet chirruping sound when either talking to you or one of their wives.

Hen turkeys, however, have a large vocabulary (and I'm indebted to Janice Houghton-Wallace for the following):

High-pitched weeing trill: It's a predator, run and hide NOW!
Indignant two-tone click: What did you just say?
Indignant purr: Don't get too close.
Rapid click: Don't get lost – you're going too far.
Loud low, high, low low peep: I am lost – help!
Variety of clicks: That was a stupid thing to do (to the one who was lost) but welcome home.
Chock chock: I'm going to teach you a lesson.
Little trilly purr: I'm off to bed.
Quack quack: I'm ready for the stag of my dreams.

A turkey can run at speeds of up to 20 mph and fly at 55 mph.

Housing and fencing

Turkeys are hardy birds and excellent flyers – given the chance they will roost high up in trees, but they are also vulnerable to foxes so if you live in an area with a high density of these mammals you may wish to contain your birds. Clipping one wing will prevent them flying far but still allow them to hop up on to perches.

They will need quite a large shed, barn or stable with perches around 60 cm (2 feet) off the ground. Remember that they have large tails so do not site your perches too close to a wall. These should be poles of a minimum of 5 cm (2 in) wide. In a large barn the perches could be higher, or the birds might even go up into the rafters as long as there is enough space for them to fly down, but in an enclosed area they must be low so that the birds don't injure themselves when they hop down to the floor.

Ventilation is important and shavings on the floor are preferable to straw. It is a good idea to provide enticing nest boxes if your birds are in any way free range, to dissuade them from laying out, which they will do for preference – a sure way for a fox to find them. A tea chest is about the right size, and they will appreciate a roof as they are searching for somewhere exclusive and hidden. Even an old tyre filled with hay and placed in a dark corner could be used.

Bubbly jock –
Scots vernacular for turkey.

Your birds may become well behaved and put themselves to bed, but it may be the case that you have to herd them in every evening – something that should be taken into account when you are considering becoming a turkey keeper.

Fences will need to be 1.8 m (6 feet) high to contain your birds, even if they have one wing clipped, as they can hop to at least 1.2 m (4 feet), but will need to be higher to deter foxes. Stags in side-by-side pens will still fight through the wire and can destroy ordinary chicken wire. Specially made electric netting is available, which is ideal as long as the electricity supply is checked every day and nothing is allowed to short it, e.g. weeds or twigs fallen from trees.

Feeding and water

Turkeys must have fresh clean water at all times – small buckets such as those used for feeding calves are ideal, as the birds must be able to get their heads in and reach the water. Hygiene is also very important and the more birds you have the more care you must take.

As always, feeding is not an exact science and amounts will depend entirely upon whether your birds are totally enclosed, totally free range or a mix of the two. As a very vague rule of thumb feed each adult bird a double handful of pellets in the morning and a single handful of mixed corn in the evening. The best way is to measure out the number of handfuls for the number of birds kept and find a tin or bowl of the right size. Put the pellets in a container on the ground – an old roasting tin or such like – then watch and see whether the birds gobble them straight up or leave some. If any pellets are left after 20 minutes then you are feeding too much and any food left lying around will attract rats and other vermin. If your birds are permanently housed then acquire some hanging feeders that keep the food off the floor. Even if your birds are out in the day, if their house has room, then it is a good idea to feed them inside in order to keep their food dry.

Three strikes in a row at bowling is known as a 'turkey' or 'triple'.

Keep your feed in mouse- and rat-proof containers – dustbins are ideal, galvanised ones even better, or you might want to lash out on purpose-built feed bins.

Turkey feed comes in a wide variety of crumbs and pellets – read the instructions on the sack and make sure you are feeding the correct type:

❏ Turkey starter crumbs: day-old to around 5 weeks. Feed ad lib.
❏ Turkey rearer pellets: around 5 to 8 weeks. Feed ad lib.
❏ Turkey grower pellets: from around 8 weeks. (You may find you can gradually mix these pellets in with the crumbs from 5 weeks and miss out the rearer pellets but do this carefully.)
❏ Turkey finisher pellets: if rearing for the table feed these from 12 weeks onwards.
❏ Turkey breeder pellets: move from grower pellets to breeder pellets if the birds are not being reared for the table. Suitable for laying hens and all adult birds.

Turkeys also require grit in order to grind their food in their gizzards. Free-range birds may be able to find enough for themselves but will still appreciate a ready supply. Oyster shell can also be offered during the laying season and your feed merchant will probably be able to supply you with mixed bags of grit and shell. Put this in a container in the run – it won't matter if it gets wet as long as it can drain.

Breeding

Turkeys make good mothers and will happily go broody and rear their own families. The males will start to show interest, and therefore display, as soon as the days start to lengthen in early January and the hens should then be fitted with 'saddles'. These are made of either leather or canvas and fit on the hen's back with straps round the wings. The hen may feel a little odd at first but will soon get used to wearing this strange contraption and it will stop her getting damaged when the male mounts her.

Younger stags are more fertile than older males – fertility decreases after about the age of four.

straps should be webbing, not elastic.

Rope to help the stag grip.

Large hens can also be used and can sit on 8 or 9 turkey eggs. The incubation for turkeys is 28 days so she will have to be patient. Incubators are another possibility. Methods are exactly the same as described on page 15.

If using an incubator follow the manufacturer's instructions with particular regard to humidity. Turn off the rotator on day 25. Hatching may take up to 2 days and this is normal. Transfer to your brooding unit (see page 17) and feed the poults turkey starter crumbs ad lib – your local feed merchant may not have these in stock at all times so think ahead and order some. Do not feed them chick crumbs as these contain an anti-coccidian drug which could kill the poults.

Turkey feathers

Five fat turkeys are we. We slept all night in a tree.
When the cook came around we couldn't be found
so that's why we're here, you see.
Children's traditional Thanksgiving song

If using a broody hen then add a handful of mixed corn daily for her. You could use a combined method – hatching the eggs in an incubator and then if you have a broody, introducing the poults to her and she will do the rest. A degree of tact is required for this method.

First move the broody to wherever you want to keep the poults – let her sit on some china eggs and then when all the poults have hatched and after dark, gently place them one by one under the hen, removing the eggs as you do so. As long as she fluffs out her feathers and murmurs encouragingly to the poults, all will be well. She may not take kindly to them, however, so a heat lamp should always be to hand.

Bourbon Red poults

At around 5 weeks your poults will no longer need heat during the day and can venture outside. They will still be very small so should be contained and protected from predators – even birds such as magpies might be a problem and it is a good idea to make them a small cage with netting over the top. With a broody of course they can step outside immediately as she will be there to brood them if they get chilled and will also protect them from aerial attackers. Depending on the time of year and weather gradually raise your heat lamp until after week 6 they should be hardy enough to be 'off heat'.

You won't be able to tell the sex of your birds until they are at least 10 or 12 weeks old. The hens have smaller and paler heads and are more vocal. Males are usually taller than hens and the caruncles on their head and neck will be more obvious. It is only the male that gobbles – so once they start this is another sure sign.

Parts of a turkey

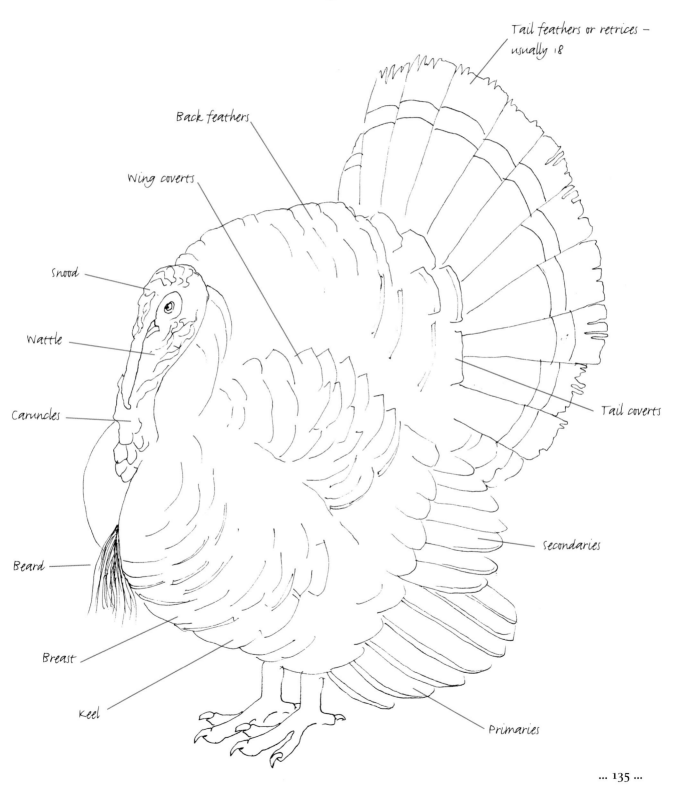

Back feathers

Wing coverts

Tail feathers or retrices – usually 18

Snood

Wattle

Caruncles

Tail coverts

Secondaries

Beard

Breast

Keel

Primaries

Blue

Blue in plumage but not in temperament

The important thing about this breed is that the colour should be even all over. The female can be a slightly lighter shade than the male and both can be light or dark. There is also a very pale version known as lavender that, although popular, is affected by a lethal gene that can cause blindness, most particularly in the females, but this need not be a problem if breeding stock is chosen with care.

Blue hen

Blue poult

ORIGIN	CLASS	POULT COLOUR	
USA	Light	Yellowish-white with blue tinge	
	YOUNG WEIGHT		**MATURE WEIGHT**
STAG	7.3–10.4 kg 16–23 lb		10.0–12.7 kg 22–28 lb
HEN	3.6–6.3 kg 8–14 lb		5.4–8.1 kg 12–18 lb

Blue stag

Bourbon Red

A popular possessor of bright chestnut plumage

The Bourbon Red is an old American variety created by crossing Buff, Bronze and White Holland and named after Bourbon County in Kentucky where it was developed. The plumage is a rich chestnut red with black lacing in the male and white flight feathers and tail in both sexes.

The Bourbon is a large bird, particularly popular with smaller producers thanks to its foraging skills and placid character.

Bourbon Red hen and poults

ORIGIN	CLASS	POULT COLOUR	
USA	Heavy	Head light red with dark spot at back. Body yellowish-brown with dark stripes.	
	YOUNG WEIGHT		MATURE WEIGHT
STAG	10.4 kg 23 lb		14.9 kg 33 lb
HEN	6.3 kg 14 lb		8.1 kg 18 lb

Bourbon Red stag

Bronze

A popular choice with a glint in its feathers

The Bronze gets its name from the magnificent plumage that reflects light and appears a metallic greenish-bronze. The breed originated in the eighteenth century, the result of crossing domesticated European turkeys brought by colonists with the tame wild birds that they found on their arrival in America. The name 'Bronze' was formally adopted in the 1830s and recognised by the American Poultry Association in 1874. During the twentieth century, the Bronze was crossed with fast-growing commercial birds to create the Broad Breasted Bronze. This became the standard turkey used by the meat industry until the mid-1960s, but its massive breast made it unable to mate naturally.

Today the hardy, slower-growing, naturally mating standard Bronze is once again appreciated for its dense-textured, flavourful meat and is favoured by many organic producers. Bronze turkeys need plenty of space to move around in and grow. They thrive in sheltered environments such as orchards and benefit from being able to graze naturally.

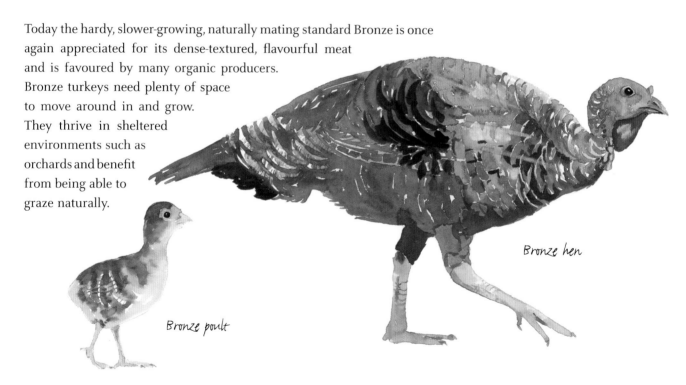

Bronze hen

Bronze poult

ORIGIN	CLASS	POULT COLOUR	
Europe	Heavy	Light brown head with dark spots – back dark streaks. Yellowish-white underneath.	
	YOUNG WEIGHT		MATURE WEIGHT
STAG	11.3–15.9 kg 25–35 lb		13.6–18.1 kg 30–40 lb
HEN	6.4–10.0 kg 14–22 lb		8.2–11.8 kg 18–26 lb

Bronze stag

Buff

Handsome is as handsome does

The Buff was once a popular turkey in the mid-Atlantic states of the USA, until the Bourbon Red took over, and although entered in the APA Standard of Perfection in 1874 was dropped in 1915. It retained its popularity in the UK and is still recognised by the British Poultry Association.

The buff colouring is described as 'cinnamon' and should have no black or white marking on any of the feathers, although the primaries and secondaries can be very pale. Good examples of this vigorous and charming breed are now rare.

Buff poult

Buff hen

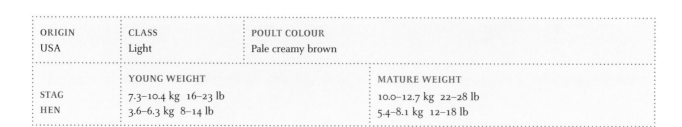

ORIGIN	CLASS	POULT COLOUR	
USA	Light	Pale creamy brown	
	YOUNG WEIGHT		**MATURE WEIGHT**
STAG	7.3–10.4 kg 16–23 lb		10.0–12.7 kg 22–28 lb
HEN	3.6–6.3 kg 8–14 lb		5.4–8.1 kg 12–18 lb

Buff stag

Narragansett

An ancient breed with a delightful nature

Narragansett Bay in Rhode Island is where this breed was developed in around 1700, supposedly by crossing Norfolk Blacks from England with the Eastern wild turkey. The plumage is similar in colouring to the Bronze but the bronze colour is replaced with steel grey edged with black and should lack the bronze or greenish sheen of the Bronze.

This old breed was once very popular, particularly in New England where its meat quality and placid temperament were admired, but it is now rare and only kept by turkey enthusiasts.

Narragansett hen

Narragansett poults

ORIGIN	CLASS	POULT COLOUR	
USA	Heavy	Head yellowish-grey with three dark streaks that continue down neck and body. Rest of body light greyish-brown mottled with dark brown.	

	YOUNG WEIGHT		MATURE WEIGHT
STAG	10.4 kg 23 lb		14.9 kg 33 lb
HEN	6.3 kg 14 lb		8.1 kg 18 lb

Narragansett stag

Norfolk Black

A striking bird black from head to toe

One of the first to be developed from imported American turkeys in the early 1500s this breed was then returned to America with European colonists and further crossed with Eastern wild turkeys providing the foundation stock for several other breeds, including the Bronze and Narragansett.

The plumage should be totally black with no hint of bronze creeping in. The beak is black and even the eyes are such a dark brown they appear black. The legs too should be black but as the bird ages they gradually become pinker. Simply known as Black in the USA, there is also a variety called Spanish Black, although there is some debate as to whether this is a separate breed or simply a different strain – it is very similar but can appear more upright than the Norfolks.

Norfolk Black hen

Norfolk Black poult

ORIGIN	CLASS	POULT COLOUR	
UK	Light	Black but creamy white face	
	YOUNG WEIGHT		MATURE WEIGHT
STAG	8.2–10.0 kg 18–22 lb		11.4 kg 25 lb
HEN	5.0–5.9 kg 11–13 lb		5.9–6.8 kg 13–15 lb

Norfolk Black stag

Pied

An ornamental beauty with spectacular plumage

There are many variations of black-and-white turkeys including Crollwitzer from Germany, Pied Ronquières from Belgium and Royal Palm in the USA – but what they all have in common is black-and-white banded plumage. When the stag displays the black band should be a consistent arc around the tail, and there can be few more spectacular sights.

This is a prime exhibition bird and although smaller than some breeds produces a good carcass and is an excellent choice for a smallholder, being an alert forager.

Pied hen

Pied poult

ORIGIN	CLASS	POULT COLOUR		
Europe	Light	Yellowish-white		

		YOUNG WEIGHT	MATURE WEIGHT	
STAG		7.2 kg 16 lb	9–10 kg 20–22 lb	
HEN		4.5 kg 10 lb	5.4 kg 12 lb	

Pied stag

Slate

A tricky one to breed true

Rather similar in appearance to the Blue, the Slate is also a greyish-blue but the plumage may be dotted or spattered with black. There should be no solid black feathers or any that contain brown or buff.

The Slate is thought to have been formed by crossing Norfolk Blacks with either the White or Easter wild turkey. It is very difficult to breed a show-class bird as offspring may hatch black, blue or slate.

slate poult

slate stag

ORIGIN	CLASS	POULT COLOUR	
USA	Light	Yellowish-white with blue tinge	

	YOUNG WEIGHT		MATURE WEIGHT
STAG	7.3–10.4 kg 16–23 lb		10.0–12.7 kg 22–28 lb
HEN	3.6–6.3 kg 8–14 lb		5.4–8.1 kg 12–18 lb

slate hen

White

A spectacular bird with pure white plumage

White turkeys have been around since the time of the Aztecs and are thought to be a sport of other breeds where the white gene has become dominant. There are several white breeds: British White, Austrian White, White Holland and Beltsville Small White, which all have pure white plumage in both sexes. The stag's beard is black and both sexes have pink legs and blue-black eyes.

This turkey is beloved of commercial producers thanks to its white pinfeathers that make for a clean plucked carcass.

White stag

White poults

ORIGIN	CLASS	POULT COLOUR		
USA	Light	Pure white		

	YOUNG WEIGHT		MATURE WEIGHT	
STAG	7.3–10.4 kg 16–23 lb		10.0–12.7 kg 22–28 lb	
HEN	3.6–6.3 kg 8–14 lb		5.4–8.1 kg 12–18 lb	

White hen

Common ailments

Most of the ailments detailed below are only encountered if your birds are kept in overcrowded or dirty conditions – as a rule all the domestic fowl mentioned in this book are hardy and will be doing their very best to maintain their feathers and stay healthy.

PROBLEM	CURE	D	Go	Gu	P	Q	T
Aspergillosis: mould from fungi growing on damp bedding causes lung congestion.	Fungal treatment for birds. Keep bedding dry and clean.	●	●				
Avian influenza: H5N1 breeds in respiratory and intestinal tracts and is transmitted bird to bird. Difficulty breathing and sudden death.	Vaccination – if diagnosed mandatory slaughter of flock.	●	●	●	●	●	●
Blackhead (histomoniasis): parasite in the liver. Birds look ill and discharge bright yellow diarrhoea. They can die within 2 to 10 days.	Emtryl (Dimetridazole) has been withdrawn by European Commission but can be obtained in small quantities for pigeons or ask your vet.					●	●
Bumblefoot: caused by birds jumping down from roosts or perches or from small wound. Swelling will appear and cause lameness.	Administer penicillin or simply leave alone and it will get better on its own.		●		●	●	●
Coccidiosis: parasite in the gut. Affects chicks and poults. Birds look ill and discharge white diarrhoea and sometimes blood – gradually weaken and die.	Clean house and spray disinfectant. Treat with water-soluble coccidiostat.	●	●	●	●		●

PROBLEM	CURE	D	Go	Gu	P	Q	T
Lice: scratching and moping.	Check vent area for yellow louse and white clusters of eggs on the base of feathers. Remove eggs and sprinkle louse powder. Add powder to nest boxes.	●	●	●	●	●	●
Newcastle disease (fowl pest): this is a notifiable disease but it is very unlikely you will encounter it unless there is a national outbreak. Birds twist their necks around, are floppy and unable to stand and there may be high mortality.	There is none and there will be a mandatory slaughter programme. Ducks and geese can suffer from a different type of this disease to hens.	●	●	●	●	●	●
Red spider mite: live and breed in housing, on perches and in nooks and crannies.	Anti-red-mite spray in housing.	●	●		●	●	●
Scaly leg: scales lifted on the leg, leg looks crusty. Caused by a mite that lives under the scales.	Proprietary scaly leg topical spray.				●	●	●
Wet feather: birds look damp and unkempt because the feathers are not repelling water; caused by a lack of oil in the preen gland. May be because of poor food or muddy surroundings and a lack of clean water to bathe in.	Supply adequate water and feed plenty of greens and whole wheat.	●	●				
Worms: birds lose weight and appear listless. Green droppings. Several types: round, tape or caecal worms in the caecal tract and intestines and gape which lives in the windpipe and causes birds to gasp or gape.	Flubenvet – a powder that is added to food or water. Worm twice a year to prevent.	●	●	●	●	●	●

Useful websites

www.amerpoultryassn.com

Website of the American Poultry Association, which aims 'to promote and protect the standard-bred poultry in all its phases'.

www.defra.gov.uk

Department for Environment, Food and Rural Affairs – very comprehensive site covering all marketing rules and regulations of the UK egg and poultry industry. Large section on poultry welfare and updates on avian influenza.

www.domestic-waterfowl.co.uk

Website of the Domestic Waterfowl Club of Great Britain – contains information, breeding facts, pictures and information on where to buy quality pure-bred ducks and geese along with helpful advice.

www.feathersite.com

Comprehensive poultry site with heaps of photos and information.

www.poultrykeeper.com

This is a hobby site run by a small group of poultry-keeping enthusiasts. It contains over 500 articles and photos and lots of useful resources on keeping chickens, ducks, geese, turkeys, guineafowl and other poultry.

www.poultryclub.org

Website of The Poultry Club of Great Britain. Provides lists of breed clubs and societies with contact details, as well as a great deal of useful advice.

www.rbst.org.uk

Website of the the Rare Breeds Survival Trust, a UK charity founded to protect all native breeds of farm animals, including poultry.

www.turkeyclub.org.uk

Website of the Turkey Club UK set up in 2001 to promote standard varieties of turkey – a helpful site with lots of advice and information.

www.waterfowl.org.uk

Website of the British Waterfowl Association – oversees waterfowl standards and produces the definitive book on waterfowl.

Acknowledgements

I would like to thank the following people who kindly allowed me to use their birds as models for my illustrations or supplied me with photos to work from:

Aden Pearson: Pomeranian

Alison Keswick: Saxony

Anthony Stanway: Blue Swedish

Ashton Waterfowl: Brecon Buff

Camilla Meheran: West of England

Carole Tucker, Dave Walsh: Embden

Cecily Day: Sebastopol back view

Diana Elliott: White peahen, Indian peacock, Bronze turkey hen and Indian Runner

Elizabeth Barrett: Muscovy on water

Gillian Dixon: Pilgrim

Hillview Exotics: American Buff

Holderread Farm: Welsh Harlequin

Jane Perowne: Aylesbury and Khaki Campbell

Jeff and Nancy Smith of Cacklehatchery.com: Welsh Harlequin, guineafowl keets and peafowl chicks

Jim Champion: Peahen

Jim Williams: Bali on water

John Metzer: Cayuga duck and African gosling

John Richards: Grey Back

Kim Brook: Muscovy duck and Sebastopol gosling

Linda Dick: Steinbacher

Majestic Waterfowl Sanctuary: Sebastopol

Mark Walters: Harlequin quail

Mike Sumner: Buff turkey

Niels Walker, Rosie Yells: Norfolk Black

Porter Rare Breeds: Narragansett

Rick Swartzentrover: Orpington

Robert Stephenson: Bali duck, Czech, American Buff and Brecon Buff geese

Sarah Barrett of Parkside Quails: quail

Tim Daniels: African, Chinese and Brecon Buff

Tom Davico: Rouen

Tome Davies: Rouen drake

Walt Leonard: Black East Indian

Ziemek Tas: Khaki Campbell duckling

I would also like to particularly thank Nino Castellano and Sarah Barrett for help with quail; Janice Houghton Wallace for advice on turkeys; my sister, Carol Keefer, and great-niece Cecily Day for photographing American breeds; and Mick and Bunny Newth of the Atelier Montmiral for introducing me to drypoint.

Glossary

autosexing: a breed in which it is possible to tell the sex on hatching by colour

bean: the small round bump on the end of a duck's bill

beard: in a turkey, the plume of hairs that protrudes from the breast of a stag, which may be up to 15 cm (6 inches) long

candling: looking through the shell of an egg with the aid of a strong light to determine whether the egg is fertile

caruncles: fleshy protruberances on a turkey's head and neck that can change colour; also on the head and neck of a Muscovy duck

cilium: eye ring

coccidiostat: drug put in feed to prevent growth of parasitic coccidia

crop: the sac-like organ in which food is accumulated before it passes to the gizzard

dewlap: fold of skin below the lower mandible on geese

gizzard: the muscle where food is ground up

lobe: in waterfowl, the paunch in between a bird's legs. Dual-lobed indicates two paunches; single-lobed one paunch.

moult: annual replacement of feathers

pipping: when the unhatched gosling, duckling, poult or chick starts to break through the shell in order to hatch

poult: a young turkey up to 16 weeks

saddle: leather or canvas pad placed on a turkey hen to protect her from damage from the stag during mating

set: to put eggs under a broody or in an incubator to hatch

snood: fleshy muscle on the face above the beak that extends when the stag displays and contracts when not displaying; hens have very small snoods above the beak that do not extend

spur: bony protrusion on the leg, above the foot

standards: very precise description of a breed or bird that should be consulted when showing

wattle: fleshy flap at the front of a turkey's neck, surrounded by caruncles or fleshy protrusions; close to the beak of a guineafowl

Index

Page numbers in *italics* refer to illustrations. Page numbers in **bold** refer to glossary terms.